Watershed Management

Field Manual:

Watershed survey and Planning

by

T.C. Sheng

Food and Agriculture
Organization of the United Nations

United Book Prints

Reprinted in India by:

UNITED BOOK PRINTS
(Imprint of Scientific Publishers, INDIA)
4806/24, Ansari Road, Daryaganj,
New Delhi 110 002, INDIA
E-mail: info@scientificpub.com
Website : www.scientificpub.com

Print : 2024

ISBN: 978-93-88399-63-0
eISBN: 978-93-88399-64-7

Printed in India

TABLE OF CONTENTS

LIST OF FIGURES

LIST OF EXAMPLES

FOREWORD

The FAO Watershed Management Field Manual Is published within the FAO Conservation Guide Series as Conservation Guide W13 and will consist of seven separate volumes:

Watershed Survey and Planning (this Volume)

Slope Treatment Measures and Practices

Vegetative and Soil Treatment Practices

Road Design and Construction in Sensitive Watersheds

Landslide Prevention Measures

Gully Control

Water Harvesting

The objective of this Manual is to assist professionals concerned with the planning and implementation of watershed management activities by providing practical information supported by examples from a wide variety of situations. However, the watershed situation in each country is unique and h is impossible to provide step by step solutions which will apply In any specific area. The user of the Manual is therefore, Invited to add the local experience and information deemed necessary.

We wish to acknowledge the valuable work of Mr T.C. Sheng, Professor. Department of Earth Resources. Colorado State University on which this Volume was based. Mr S. Dembner, FAO Consultant, edited the first draft on which many useful comments and suggestions were received from the land and Water Development division of FAO's Agriculture Department and from the Policy and Planning Service of the Forestry Department. It has been prepared for publication by Mr LS. Botero and Mr T. Michaelsen of the Forest and Widlands Conservation Branch. It Is hoped that the publication will be useful to all orofessionals Involved in watershed management activities.

ABSTRACT

This document is a guide for watershed managers and planners, providing basic knowledge and practical approaches for the survey and planning of small upland watersheds with combinations of forest, cultivated and grazing lands and populated mostly by small farmers. The manual consists of three pans. The first part gives an overall Introduction and explains what preparatory work is needed. The second parts covers survey approaches and techniques together with specific examples. The third part describes planning and plan formulation

INTRODUCTION

This manual is intended to provide basic knowledge and practical approaches to watershed managers and planners, who are involved in watershed survey and planning in developing countries.

The watersheds in mind are upland or hilly watersheds with combinations of forest, cultivated, or grazing lands and populated mostly by subsistence farmers.

The guide is aimed primarily at middle-level technicians in order to enable them to carry out the actual planning work, with the assumption that concepts, strategies, problems and overall solutions have already been worked out at the appropriate decision making levels in consultation with the land users.

A watershed programme, in order to be effective, has to carry out a relatively intensive effort in a limited area, which therefore has to be very carefully identified and selected. The important catchments of a country will typically be mountain areas with high rainfall and their priority rating will depend on the related downstream interests, city water supply, hydroelectric power generation, irrigation schemes, floodplain protection, etc. They will in many cases not coincide with priority areas for national soil conservation programmes aiming at increased sustainable food production from the nation's farm land, or community forestry programmes designed to meet the needs of the local people in areas of scarcity of wood products.

Watershed problem analysis, incentive packages, road construction standards, enforcement of forest protection regulations, etc. are therefore not necessarily those applicable to rural development, soil conservation, and forest protection, etc. nation-wide. However, any special arrangement made in a critical watershed places severe limitations on its applicability and sustainability outside the catchment and over time.

The watershed planner should be aware therefore of new approaches and strategies in soil conservation and land management, farming systems analysis and development, community forestry, etc. and realize that although the identification and selection of priority watersheds is made at central level with downstream interests in mind, it is the upstream land user who is effectively the watershed manager through his or her land use decisions. The challenge is to harmonize "top-down" central planning selection of critical and priority watersheds with a "bottom-up" approach in which farmers will participate actively because of the benefits to them from improved land husbandry practices.

The survey examples given in this manual are for reference only and should not be taken as step by step instructions. Watershed management is site specific and no method, however sound, can be applied universally without modifying it to suit local needs.

The manual is divided into three parts. The first part gives an overall introduction and explains what preparatory work is needed. The second part starts with survey approaches and techniques in general, followed by a long section (chapter 7) of specific techniques and examples, many of which are from actual experiences. The third part concentrates on planning and plan formulation. A brief introduction to the use of microcomputers for this purpose is placed in Appendix 3 for reference.

It should be pointed out that, in practice, survey and planning is a continuous process and that the divisions included in this manual are only for the sake of analysis and clarity of presentation.

1. INTRODUCTION AND CONCEPTS

1.1 The need for watershed survey and planning

The protection, improvement and rehabilitation of mountain and/or upland watersheds are of critical importance in the achievement of overall development goals. Recognizing this, many developing countries are turning increasing attention and resources to the field of watershed management. Initial efforts have often been "fire-fighting" in nature, i.e. an immediate but isolated response to a perceived problem. However, in many if not most developing countries, the nature and magnitude of the problem of watershed degradation, and the scarce availability of resources mandate a comprehensive long-term approach. The keys to successful implementation of any such effort are accurate and appropriate survey and planning.

1.2 Definitions

As a starting point, it is appropriate to set some basic definitions. A watershed is a topographically delineated area that is drained by a stream system, i.e. the total land area that is drained to some point on a stream or river. A watershed is a hydrological unit that has been described and used as a physical-biological unit and also, on many occasions, as a socio-economic-political unit for planning and management of natural resources. Catchment is often used as a synonym for watershed. There is no definite size for a watershed; it may be as large as several thousand square kilometres or as small as only a few square kilometres.

A watershed is differentiated from a river basin in that a river basin, with its trunk stream flowing to the sea, may encompass hundreds of watersheds and many other types of land formations.

Watershed degradation: Watershed degradation is the loss of value over time, including the productive potential of land and water, accompanied by marked changes in the hydrological behaviour of a river system resulting in inferior quality, quantity and timing of waterflow. Watershed degradation results from the interaction of physiographic features, climate and poor land use (indiscriminate deforestation, inappropriate cultivation, disturbance of soils and slopes by mining, the movement of animals, road construction, and badly controlled diversion, storage, transportation and use of water). Watershed degradation, in turn, leads to accelerated ecological degeneration, reduced economic opportunities and increased social problems.

Watershed management: Watershed management is the process of formulating and carrying out a course of action involving the manipulation of resources in a watershed to provide goods and services without adversely affecting the soil and water base. Usually, watershed management must consider the social, economic and institutional factors operating within and outside the watershed area.

All watersheds contain many kinds of natural resources - soil, water, forest, rangeland, wildlife, minerals, etc. In developing and managing a watershed, the use of some natural resources will be complementary while others will be competitive. For instance, logging may affect water resources and recreation. Changing intensive land use to less intensive ones may benefit soil and water resources. The key is to use these resources as efficiently and perpetually as possible, with minimum disturbance to the watershed as a whole. Although in many cases, watershed managers may not be the decision-makers on resource uses, their task is to plan and carry out practices which will encourage those uses which are complementary and suggest preventive and protective measures for those uses which could impair the watershed.

Since watershed management involves decision-making about use of resources for many purposes, a multi-disciplinary approach is essential. Work should include government institutions from various disciplines, and also involve people from different parts of society. On the other hand, involvement of too many elements in planning and decision-making can lead to inefficiency and unsatisfactory end results. Participation should be limited to representatives from key government institutions and the local communities which will be directly affected. For example, if the main objective is to protect an irrigation reservoir, the irrigation agency should be involved and not the power company. However, its inclusion might well be necessary if the reservoir was to be used as a source of hydro-electric power.

Watershed management is an ongoing undertaking. New elements, both man-made (road building, mining, logging, and cultivation) and natural occurrences (landslides, wildfire, floods) may become a factor at any time. It is important to remember that when new challenges arise, the original management plan must be revised. It is the watershed manager's or planner's responsibility to make the government authorities aware that watershed management is a continuous and flexible process.

Watershed survey and planning: Watershed survey and planning is the preparatory work which, if properly conceptualized and carried out, permits the successful implementation of actual watershed management.

1.3 Four levels of survey and planning

Watershed survey and planning should be undertaken at four levels with a problem-oriented approach (See Fig. 1).

National level

At the national level, a quick reconnaissance type of survey, assisted by aerial photographs or other remote sensing techniques, is often sufficient for identifying major watershed problems and areas. This type of survey can identify broad land use categories, main causes of disturbance and, combined with existing data, can provide enough information for a simple classification of the nation's watersheds.

Fig. 1

VARIOUS LEVELS OF SURVEY AND PLANNING

-With Emphasis on Problem-Oriented Approaches-

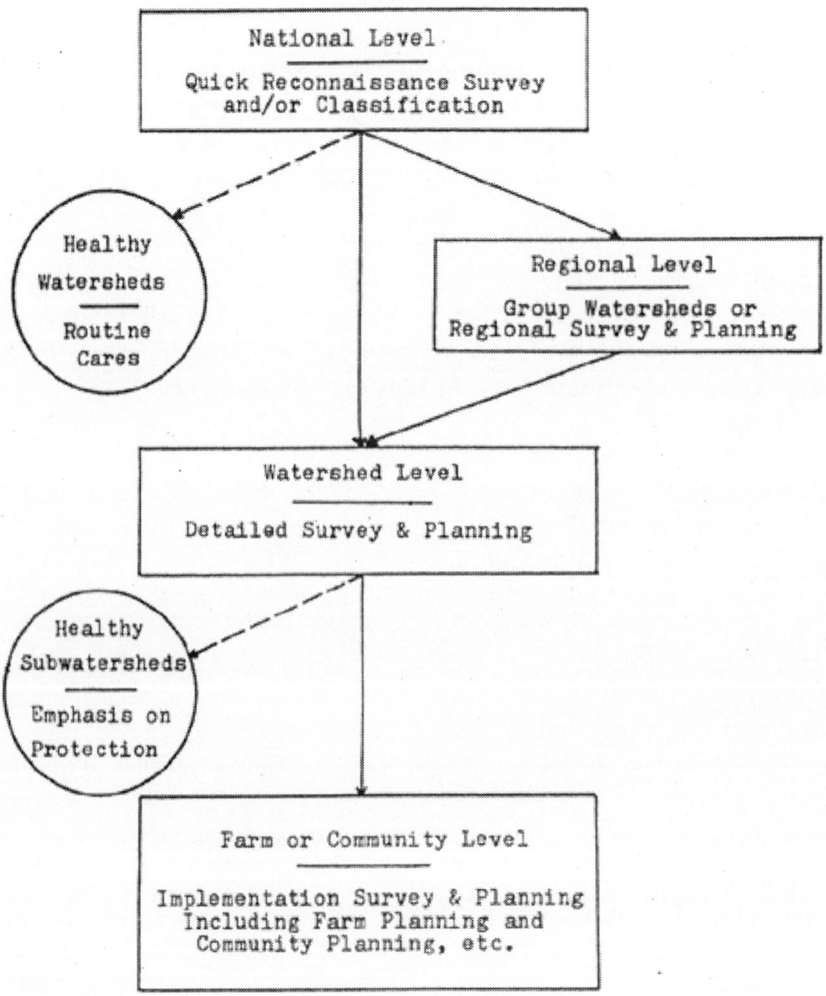

The main purpose of this overall classification is to identify the following important items:

- nature of watersheds, i.e. municipal watersheds, forest and wildland watersheds, agricultural watersheds, etc.;

- main problems and critical areas, i.e. problems caused by man, nature or both, seriousness of the problems, extent of critical areas, etc.;

- watershed sites, i.e. upland or highland watersheds, lowland watersheds, watersheds with downstream interest, watersheds without downstream interest, etc.

This classification can be an important input in determining national policy and in setting priorities.

Regional or district level

Regional or district level survey and planning is either carried out specifically for a cluster of watersheds or in conjunction with regional development plans. The work covers a more restricted area than the national study but is not necessarily as detailed as the plans for individual watersheds. These studies are important in the formulation of long-term development plans for the region or district.

Watershed and sub-watershed level

Most detailed survey and planning is carried out on the watershed level, both because a watershed is a functional unit which links upstream and downstream areas in an integral system, and because it is a convenient unit for planning and economic analysis. This manual aims at this level. In a large watershed, detailed survey and planning can also be concentrated on sub-watersheds with particularly serious problem areas or critical areas.

Farm or community level

Individual farm planning, group farm planning and planning for community development are also necessary. These can either be done during the planning period or at the beginning of the implementing stage, depending upon actual needs. The main objective is to improve farm management and community development within the watershed area. Emphasis is usually put on conservation as well as on development.

Local survey and planning provides basic data and also involves local farmers and communities in the planning process. If there are local bodies such as local government, farmers' associations, private interest groups, etc., their representatives should be involved in the planning process. Direct survey of farmers or watershed inhabitants by properly designed questionnaires on important watershed issues is also a way to get local people involved.

1.4 A problem-solving approach

To make use of limited manpower, resources and time, watershed survey and planning should be carried out in as practical a manner as possible. Surveys should be oriented towards identifying main objectives and major problems, and plans and recommendations should be centred on solving or alleviating these problems, although the overall potential of the watershed should not be neglected.

Major watershed problems

Major watershed problems vary from country to country, but the following list identifies some of those most common to developing countries. Most of these are interrelated and cannot easily be separated for diagnosis. However, for the sake of presentation they may be grouped as follows:

- *Socio-economic*

Rural poverty in the uplands, causing migration to crowded urban centres and/or destroying watershed resources.

Improper land use (slopeland farming, shifting cultivation without proper fallow, overgrazing, etc.) resulting in degradation of land and other watershed resources.

Deforestation, thereby increasing hazards of seasonal flooding and/or drought downstream.

- *Technical/Institutional*

Poorly planned and executed development activities (roads, housing, mining, recreation, etc.), impairing streams and polluting the natural environment.

- *Natural*

Natural disasters (heavy storms, landslides, wildfire, etc.) damaging watershed conditions.

Natural and accelerated soil erosion, causing heavy deposits of sediment in storage reservoirs, irrigation channels and other public installations.

Resource availabilities and constraints

Practical watershed planning should not overlook resource availabilities and constraints. An over-ambitious or unrealistic plan is less likely to be approved or implemented successfully.

Some of the main constraints facing watershed projects in developing countries include:

- lack of funds;

- insufficient manpower, especially at the professional level;

- poor coordination among government organizations;

- low mobility and insufficiently equipped field staff;

- lack of data and research for continuous improvement;

- other socio-economic, institutional or policy constraints.

These constraints should be taken into consideration seriously and strategies to overcome them should be developed at the early planning stage. Watershed managers or planners should find out what resources are or will be available to realistically manage the watersheds.

Technological considerations

The capacity of present technology to cope with the major problems of the watershed is a subject to be well considered at the planning stage. If expertise is insufficient, technology exchanges with other countries should be contemplated. This may include sending fellows abroad, or inviting foreign consultants.

Technology transfer from government staff to local farmers is also important and should be considered at the planning stage. Proper extension, education and training activities for farmers in the watershed area will help ensure ongoing success. (See 10.3 for details).

If there are not enough professionals or technicians to handle the work of survey, planning, design and implementation, an early identification of the problem is always preferable, because both institutional strengthening and personnel training take time. Experience in many developing countries has shown that watershed management or soil conservation projects can only grow as fast as trained and experienced persons are available.

In many developing countries, there are enough agronomists, foresters and agricultural extensionists, but watershed specialists, forest engineers, hydrologists, specialists in torrent control, landslides and resource economics are often sorely lacking. Therefore, the question of how to recruit and train the needed technical personnel for upgrading the capabilities of various agencies must be taken into serious consideration.

2. PREPARATORY WORK ONE: IDENTIFICATION OF WATERSHED PROBLEMS, OBJECTIVES AND PRIORITIES

2.1 Collecting existing data

Collecting existing data is the first step toward comprehensive survey and planning of a watershed. In many countries, soil surveys, geology surveys, forest inventories and hydro-meteorologic studies may have already been carried out. Their reports, statistics and maps can be of great help in watershed survey and planning. Other government agencies may possess aerial photos, contour maps, ownership data and information of infrastructures. Spending a little time to search for them is usually worthwhile and rewarding. Avoiding a duplication of effort will save time and money. Also, historical data such as rainfall, streamflow, land use history, various development plans cannot be obtained just by field surveys, but must be extracted from the files of the appropriate government agencies.

A list of data sources should be prepared and related institutions contacted. A central file of relevant maps, reports and records should be created and routinely updated.

After the existing data have been collected and analysed, then a preliminary survey plan can be drawn up to check, add and update the existing information. By so doing, the overall survey time can be much shortened.

2.2 Quick identification of watershed problems

Before starting formal survey, preparatory investigation is often needed. The main purpose is to identify major watershed problems and collect or check preliminary information. The usual activities may include:

- collecting first-hand information on the nature and extent of physical and social watershed problems to facilitate future detailed planning;

- using available photos, maps and other data to become familiar with watershed conditions;

- examining the number, reliability and condition of existing hydro-meteorological stations in the watershed;

- interviewing local agencies, institutions, communities and farmers to obtain their views, interests and concerns about the watershed;

- arranging for future survey work and estimating survey expenses.

Physical problems

These problems are usually not difficult to detect or identify. Steep slopes, bad lands, slide-prone soils, weak geologic formations, etc. can be easily found by observation or with the assistance of existing maps. Problems such as heavy and intense rainfall, excessive run-off, torrential flows and strong winds should be identified from weather and hydrological data or by gathering information and evidence locally.

Resource use problems

Problems such as shifting cultivation, forest destruction, fire, over-grazing, poor road construction and maintenance and uncontrolled mining should be identified and, if possible, the causes should be determined. Clear identification of these problems at the preliminary stage will benefit the follow-up surveys and planning as well as the formation of a realistic policy in the future.

End problems

The final effects of watershed degradation - soil erosion, landslides, heavy sedimentation, water pollution, floods and droughts, etc. - must be identified as quickly as possible. This can be done partly by observation and spot checking and partly from data obtained from water resource agencies and local inhabitants. By reviewing or analysing existing information, the history, frequency and extent of these problems can also be determined.

Socio-economic and other problems

Serious socio-economic problems can be major obstacles in carrying out watershed work in the developing countries. Any serious problems should be identified at the beginning of the planning stage. These may include land tenure, poverty, education, low acceptance of innovations, seasonal shortages of labour, etc.

2.3 Considering management possibilities

Recognizing challenges

It is not enough to simply identify watershed problems; possible challenges must also be considered when work is to be initiated.

The major challenges facing watershed managers in developing countries are highlighted below:

- watershed projects usually deal with thousands of people and therefore have broad political implications, hence political interest. Sometimes, politicians' views may differ considerably from those of the technical experts regarding the types, priorities, and timing of watershed work;

- public goals may not always coincide with the interests of private people, farmers or watershed inhabitants. For example, farmers tend to maximize the returns from their lands as fast as possible whereas governments may wish to slow down the use rate in order to conserve resources and protect the watershed;

- a technically sound plan or the most effective treatment for watershed protection may not always be acceptable to the local communities for a variety of reasons, including requirements of high labour inputs or cost;

- many watershed projects may seem financially unattractive to government or international financing agencies due to difficulties in putting a monetary value on "intangible" benefits;

- bottom-up planning is a necessity but often difficult to do successfully, especially in watersheds populated by numerous small farmers.

There are no universal answers to these challenges. However, managers and/or planners should examine them carefully and derive proper counter measures in the context of given conditions.

Examples of effective work in the country

If there are examples of successful and effective projects in watershed management, forestry or rural development in the country, a quick study should be made of their setup, administration, management techniques, accomplishments, etc. The lessons learned will aid in the planning, design, and future implementation of the proposed watershed project.

Level and kind of management expected

Consideration should also be given to the appropriate level and kind of management a watershed will receive in the future. This will, of course, depend on the seriousness of the watershed problems, the urgency of the task, and the resources available to do the work.

Classically, watershed work can generally be divided into three categories: protection; improvement; and rehabilitation. Protection measures are employed to maintain the status quo. Improvement techniques are used to obtain water yield benefits. Watershed rehabilitation is applied to seriously deteriorated watersheds and usually requires more work, more time and more money. Unfortunately, watersheds in developing countries often require rehabilitation. Since the last decade or two, watershed development has become a new category of work. It emphasizes the development of all resources in a watershed including human resources.

These categories of work are sometimes all present in a single watershed; it is the responsibility of the planner to work out the right combinations according to the nature and extent of the problems identified.

Possible investment

Finally, investment needs in a watershed must be considered against the possible sources and amounts of funding. The need for heavy and long-term investment should be brought to the immediate attention of the government. More avenues of possible investment should also be explored. For instance, investment in production and development type of activities in conjunction with watershed conservation may attract more sources of funding.

2.4 Determining main objectives and priorities

Setting main objectives

After collecting existing data, identifying major watershed problems and considering management possibilities, the main objectives of the proposed project should then be defined.

The objectives will vary from country to country and from watershed to watershed but the following are some of the most common ones:

- to rehabilitate the watershed through proper land use and protection/conservation measures in order to minimize erosion and simultaneously increase the productivity of the land and the income of the farmers;

- to protect, improve or manage the watershed for the benefit of water resources development (domestic water supply, irrigation, hydro-power, etc.);

- to manage the watershed in order to minimize natural disasters such as flood, drought and landslides, etc.;

- to develop rural areas in the watershed for the benefit of the people and the economies of the region;

- a combination of the above.

Different objectives call for different techniques, manpower, inputs and approaches in planning. The monitoring and evaluation criteria will also be different. Therefore, main objectives should be identified and defined as early as possible.

Establishing priorities

Priority watersheds or sub-watersheds should be identified during the preparatory stage. As work cannot be carried out at the same time in all the sub-watersheds due to manpower and resource constraints, a priority list must be set.

Priorities are usually given to those sub-watersheds which are in critical condition and which are close to the main stream or to a public installation where protection is needed, e.g. a storage reservoir, water intakes or diversion dams. Many times, priority areas are also selected because of people: their enthusiasm, strategic locations, poverty or others.

Even in a priority sub-watershed, some efforts need to be started earlier than others. Therefore, a priority list of work should also be identified for future progressive planning and implementation.

Estimating survey budget

As the final stage of the preparatory work, an estimate of the needed budget for detail surveys should be prepared. Although at the early planning stage, an exact estimation is impossible, the investigators or planners must not over-estimate or under-estimate too much the needed budget.

3. PREPARATORY WORK TWO: ORGANIZING SURVEY AND PLANNING MECHANISMS

3.1 Joint planning and decentralized implementation

Survey and planning work can be organized in many ways. Unless there exists already a watershed authority or board which is capable to do planning work, experience in developing countries has shown that one of the most effective methods is "joint planning and decentralized implementation". The key philosophy of this arrangement is to respect each agency's right, jurisdiction and responsibility. When each agency understands that it has an important and autonomous role to play, the work will be carried out more faithfully and efficiently. This arrangement is better than the "staff borrowing" or "piecemeal" approach.

Agencies and groups involved

The complex nature of watershed management calls for the participation of many agencies and groups but should not involve too many. Fig. 2 shows a hypothetical list of agencies for reference.

Designation of chief responsible agency for planning

Responsibility for coordination of watershed planning activities should be delegated to a single body. Depending on the country, it can be a ministry (agriculture, natural resources, or planning), a high level authority or others. Creation of a joint committee, an inter-ministerial body or an ad hoc committee is another possibility.

In many countries, the ministry or national organization which is responsible for economic planning is designated to steer watershed planning. Designation of such a national planning organization for watershed planning facilitates coordination, ensures that watershed management is included under a national development umbrella, and helps early approval once the plan is completed. Possible disadvantages are: they may be overburdened with other planning activities; they often have limited expertise in watershed management; and they may appear authoritative and not use the same language as the field workers, thus discouraging full communication. Nevertheless, the national planning organization should normally be involved in watershed planning, whether or not it is the responsible agency.

Proper coordination mechanisms

In the developing countries, coordination is often proposed but rarely enacted due to a lack of qualified staff and resources in each agency. Therefore, not only should the coordinating agencies be well chosen, but the required duties and ways and means of coordination must be clearly spelled out in the early stages of planning. A time schedule should be agreed upon by all agencies concerned and a mechanism should be developed to check progress. Otherwise, the least efficient agencies will always control the progress of the entire project. Also, many agencies may lack funds to do extra work. If necessary, survey expense (travel, per diem, etc.) should be provided by either the responsible planning agency or the original promoting agency.

Fig. 2

A HYPOTHETIC LIST OF AGENCIES INVOLVED IN WATERSHED SURVEY AND PLANNING

-For an Irrigation Reservoir Watershed with Agricultural and Forest Lands

Agencies/Groups	*Major Functions*
Ministry of Agriculture	Convener. with overall responsibilities
Forestry Department	Member, managing forest lands
Soil Conservation Division	Member, conservation work on agricultural lands
Irrigation Authority	Member, original promoter
Farmer's Association (incld. Irriga. Association)	Member, representing local communities
Extension Service	Member, farmer's education, extension, and training
Local Government	Member, rural development and infrastructure improvement
National Planning Agency	Member, coordination and economic assessment

A steering body

Under the chief responsible agency, a steering body involving representatives of coordinating agencies and local communities should be organized for each watershed survey and planning task. The steering body's main responsibilities are:

- to coordinate all matters concerning survey and planning;

- to determine the responsibility of each participating agency;

- to establish an effective coordination mechanism;

- to acquire the necessary funds and instruments and equipment for the survey;

- to set guidelines, criteria, specifications and time schedules for the survey and planning;

- to oversee and control the quality and progress of work;

- to compile reports and a final plan or proposal;

- to implement any follow-up work required by the government, including approaching funding agencies, supervising implementation and evaluation.

Field survey and planning teams

Under the steering body or committee, a number of field survey teams should be formed according to actual needs. For the agences involved in Fig. 2, the needed teams would be:

- a land use and soil conservation team;

- a forest and natural resources protection team;

- a hydrology and engineering team;

- a socio-economic and farm management team;

- an infrastructure improvement team (if needed).

A team can have members from different agencies but its leader should be drawn from an agency responsible for the survey subject. For instance, the leader of the soil conservation team should come from the soil conservation division, although team members may come from extension services, farmers' associations and others. Each team leader reports periodically the progress to the steering committee and also serves as a bridge between his mother agency and the committee.

Liaison unit

A liaison unit or a liaison officer is sometimes needed under the steering committee to coordinate the work. This unit or officer should delegate the committee's authority. Even if the committee is terminated after the planning stage, this officer or unit can still remain to coordinate further implementation and evaluation work.

3.2 Setting guidelines and criteria

General guidelines and survey needs

General guidelines should be set out at the beginning and should include, but not be limited to:

- guidelines related to policy matters;

- statements on main objectives;

- major watershed problems as identified by the preparatory investigation;

- possible management strategies;

- guiding principles in survey and planning including priorities, time schedule and procedures, etc.

The guidelines should be short and concise. They should be discussed fully during the first steering meeting so that no misunderstandings or misinterpretations arise at later stages. In addition, specific survey needs and a scheme (see Fig. 3 for a hypothetical example) should be worked out with detailed time schedule.

Survey methodology and criteria

After the survey teams are organized, each of them should prepare a working outline explaining survey criteria, methodology and procedures. The outlines should be submitted to the steering committee for comment, discussion and approval. These outlines usually include the following items:

- sector objectives and terms of reference;

- composition of the team and coordination;

- methodology in brief;

- survey criteria including various survey forms;

-time schedule for field surveys, data analysis and reporting;

Fig. 3 A HYPOTHETIC MASTER SCHEME FOR SURVEY AND PLANNING

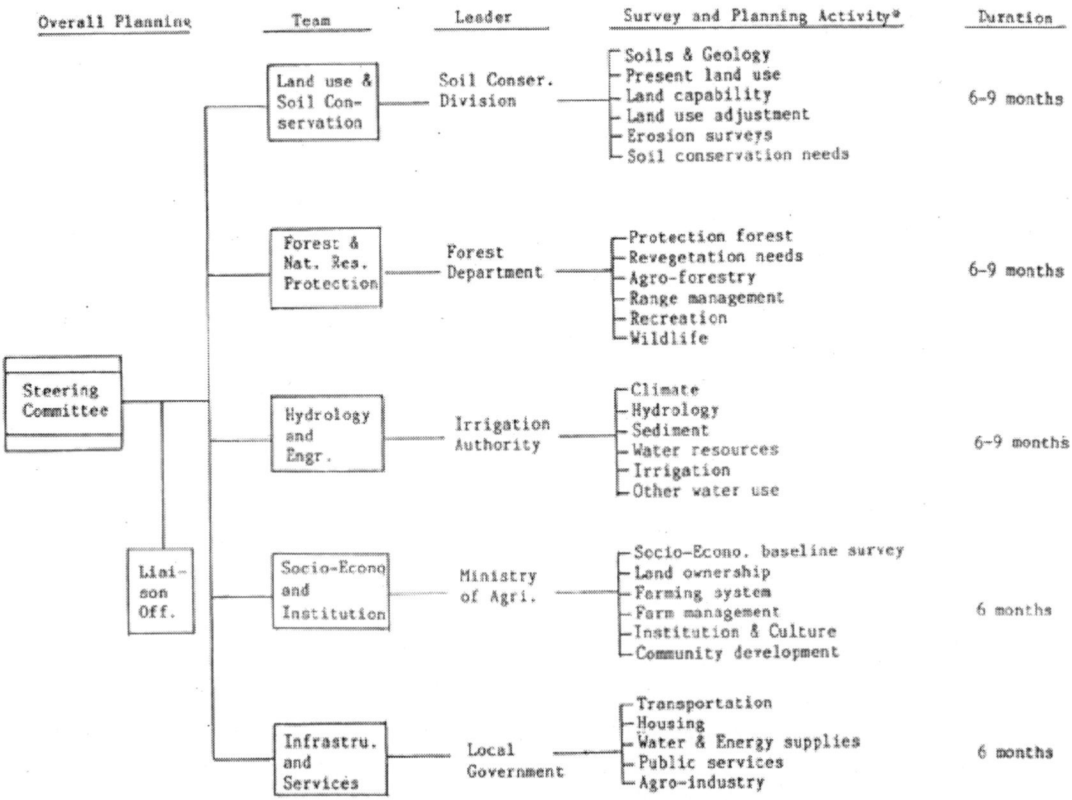

* Reduce any survey item if not necessary. Examples are given in Chapter 7.

- survey tools and transport needs;

- estimates of required funds;

- team report format.

3.3 Progress monitoring

Determining and controlling progress

Timely completion of multi-disciplinary survey and planning is difficult if there is no firm control of time and progress. Therefore, after each survey team develops its work schedule, the steering committee should establish an overall timetable. It should allow for a reasonable time period to do a satisfactory job and ensure that all work will be completed by the final deadline. An overall schedule of survey and planning activities is shown in Fig. 4. A time frame can be added to each major activity for control purpose.

Work should be scheduled to avoid conflicts; for example, no team should be kept waiting for others during the course of the survey. Important dates such as completion of field work, data presentation, submission of preliminary results and reports should be throughly discussed and determined. To ensure effective control, the committee or the liaison unit should make routine checks; any problem which hampers progress should be brought up for immediate discussion and settlement.

Periodic reporting and interim report

Completion of a watershed plan in a short time period normally requires a concentration of efforts. Periodic reporting is very necessary. Regular meetings of the steering committee should be held at least once a month and special ones should be added when necessary. Periodic reporting from each team should be concise and to the point whether oral or written.

After the preliminary data have been gathered and analysed, the committee usually needs to compile an interim report to the authorities concerned to get first reaction or feedback before going further. Changes or revisions may be needed for final reporting.

Final reporting: format and contents

The format of the final report and its contents should be decided as early as possible. Although they depend on data outcome and findings, the general format should not be determined at the last moment. While the details can be seen in Chapter 11, the following are some general rules:

- the main text should be as concise as possible. The report is mainly for administrative use and not for academic studies. An abstract or executive summary is often needed at the beginning of the report. Technical details and discussions of methodology, etc. are best put at the end or attached as appendices;

- the contents should be as practical as possible. Watershed problems should be analysed; objectives, goals, and work progress be clearly set; responsibility of each agency or sector be well defined; cost. and funding sources be identified; expected results, benefits and financial viabilities assessed; and strategies described;

- the whole report should be as illustrative as possible. Charts, simple diagrams and photos should be included to highlight important points.

Preparation for data base management

Essential data for the watershed should be well kept or stored to permit monitoring and periodic evaluation in order to detect changes and management impact over time. This can be done by using microcomputers and proper software (see Appendix 3).

Fig. 4

A SCHEDULE OF SURVEY AND PLANNING ACTIVITIES

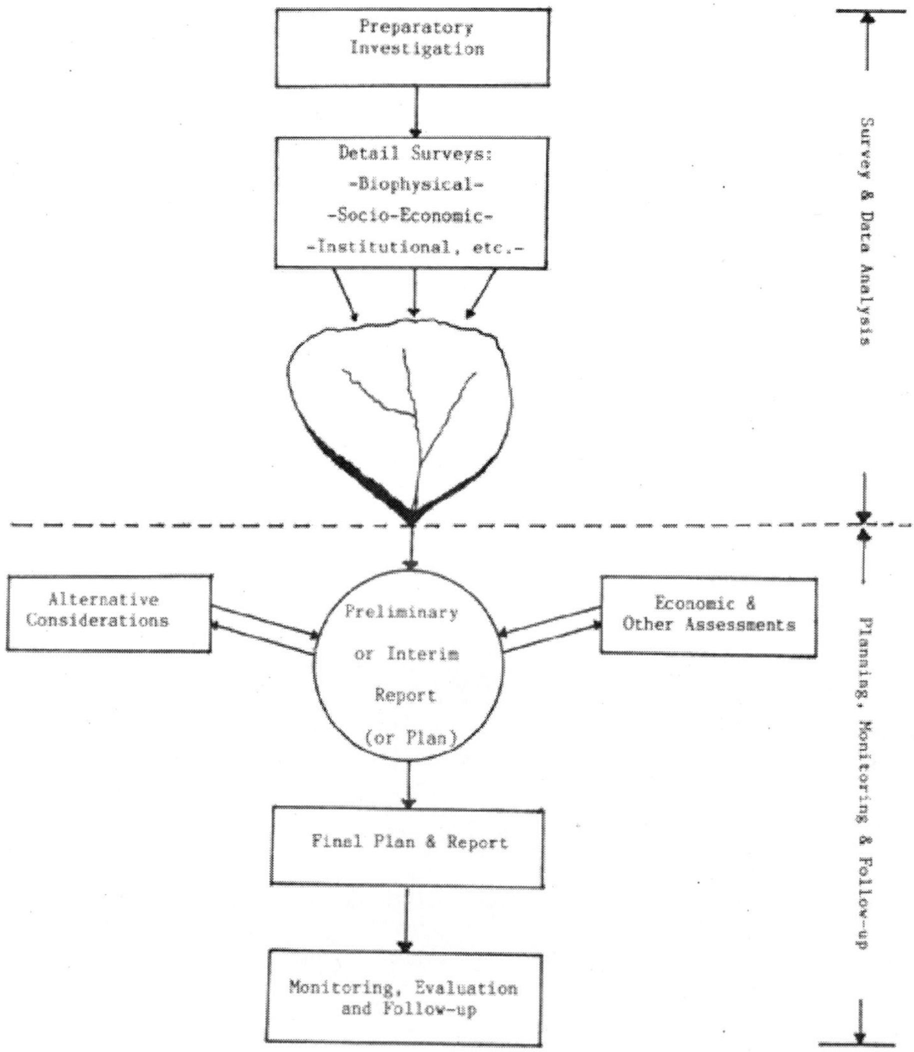

4. BIOPHYSICAL DATA COLLECTION AND ANALYSIS

4.1 Basic approaches

Problem-oriented biophysical surveys

After identifying the major problems of the watershed, detailed biophysical surveys should be designed on a problem-solving basis. Healthy parts of a watershed should be put on routine care while special attention and urgent treatment must be given to critical areas or problem sub-watersheds.

For instance, if the main purpose of the effort is to reduce sedimentation of a reservoir, the survey work should be concentrated on identifying erosion or sediment source areas. Survey of forest areas should be concentrated on identification of cut-over areas, bare areas, reforestation needs, as well as cover types, densities and hydrological conditions of the land, rather than on volume or value of timber. Detailed surveys will also need to be carried out on disturbed areas such as cultivated fields, road slopes, streambanks, mined-out areas and landslides, etc.

If the main objective is for watershed or rural development, then, the survey should concentrate on resources inventory, distribution, uses, establishment and land productivity, etc.

Collecting relevant and essential data

Collection of too much or too little data can only result in a waste of time and money. The way to avoid this is to have a careful design at the beginning of the survey. Attention should be given to what data are really needed for future management and whether they are relevant to the main objectives.

Establishing benchmarks for future surveys

Since watershed conditions change over time, future biophysical surveys will be needed every ten years or so. These periodic surveys are also used to evaluate management effects. For this reason, the initial surveys should be considered as benchmarks, and all results kept and stored for future monitoring use (see 11.3).

4.2 General guidelines

Design

Although the design of the survey will vary depending on objectives and actual needs, some general rules should be observed:

- the data to be collected should be accurate and useful for the final analysis;

- survey forms, tables and guidelines to be used in the field should be easily understood. They should be field-tested before extensive use. The forms and tables must produce objective and not subjective results;

- the survey should be designed to identify problems and their location, extent, and areas which will be useful for deciding on treatment and control measures;

- all the field surveys should be so arranged that they can be carried out orderly within the allowable time period. A network analysis or a flow chart is sometimes needed to indicate a step-by-step approach;

- to facilitate future surveys, all measurement units, mapping and photo scales, survey forms and analysis procedures and records should be kept in a standard format.

Sample size and sampling techniques

Many types of surveys do not need to cover the whole area or the whole population. However, the difficult question is how big a sample size or how many samples are considered sufficient. For practical purposes, if the known population is large, a sample of 3 to 5 percent may be adequate; if the population is small, a sample of 10 to 15 percent may be more appropriate. The sample must represent the population and allow the work to be done within the time frame and financial limits.

Many sampling techniques can be used in the biophysical survey of a watershed. The general ones are briefly described below. For those who are interested in theories and detailed techniques, standard textbooks should be consulted.

Random sampling. This technique requires a knowledge of the population to be sampled. This sampling method can be used to check large quantities of gullies, landslides, streambank cutting where total numbers can be identified (for example, from air photos) but specifications of each cannot be measured except by field sampling. This technique can also be used to investigate, for example, hydrologic soil conditions of forest lands or of the entire watershed. After sampling size is determined, a random starting point should be selected on a map. From there, the subsequent points can be decided upon by use of a predetermined distance or grid. All the chosen points should then be visited and investigated in the field. Sometimes, a random number table is used to select the samples of gullies and slides.

Stratified sampling. A population is divided into sub-populations of known size, and then random samples are taken from each stratum. For instance, survey and estimation of sheet erosion can be done by major soil types and by major land uses. In each category, random samples can further be determined for field investigation. Such strata can also be established in different elevation groups, slope categories, cover types and farm sizes to be used for various kinds of sampling surveys.

Cluster sampling. In this technique random clusters are selected and then the entire sample in the clusters is surveyed. This method is usually employed to check survival rates of tree plantations (by small parcels or rows); to investigate fuel consumption of villagers (by village), or to estimate land use patterns in different parts of a watershed.

Whatever method is used, it needs to be done faithfully and as accurately as possible. Sampling saves much time and effort compared with surveys of the entire population, but if not properly carried out, can be useless or misleading.

Mapping scales and mapping

Basic maps need to be prepared at one convenient scale. If the country's base maps are at 1:10 000 or 1:12 500 scale, the watershed maps should have the same scale in order to facilitate the transfer of information or the production of subsequent maps by superimposing one on another.

To transfer images from air photos to a map, a "Sketchmaster" can be employed. A Sketchmaster is a simple and inexpensive mapping device used extensively in many developing countries. If the map and photo scales are the same or similar, the work can be done much more quickly. For more precise mapping work, a desk type of "Map-o-graph" can be equipped to do the job. With some training, a draftsman can enlarge, reduce and transfer information from map to map.

For using computers for mapping, please see Appendix 3.

Obtaining area statistics

To obtain area statistics from maps, the dot counting method is usually employed. However, if a map contains hundreds of small parcels, the work can become tedious and time-consuming. A much simpler and quicker method which can be used by non-technical personnel has been devised recently. It is called the "cutting and weighing" method, in that parcels are coloured, cut, grouped and weighed by a balance against the weight of a known area. Various categories of areas can easily be obtained using a simple ratio calculation. The area figures thus acquired are quite accurate provided that the thickness of the paper used is constant.

When computers are used for mapping, area statistics are presented automatically and graphs are in the form of histograms or pie charts.

4.3 Data requirements

The kind of biophysical data needed for survey and planning depends on watershed problems and management objectives. Only brief descriptions are given in the following sections. Survey techniques and samples are given in Chapter 7.

General data

General data on a watershed should include watershed name, location, boundaries, size, elevation, the presence of streams, tributaries, etc. The watershed may need to be divided into many sub-watersheds and each should be assigned a number for easy identification. Further data will be needed on a sub-watershed basis. General information such as population, administrative districts, accessibility and roads, etc. is also useful.

Physical data

Data on geology and soils can usually be obtained from existing reports. Nonetheless, field checking is often necessary to verify or supplement the existing information. Geomorphological data such as drainage patterns, stream density and order, channel profiles, etc. can be obtained and/or quantified by using a good topographic map.

Data on soil hydrologic conditions are sometimes required in forest, rangeland, and cultivated lands (section 7.3 and examples 18 and 19 explain some of the methodology for reference).

Land slope information must be obtained and analysed in order to determine land capabilities of a watershed and appropriate conservation or treatment needs. With a topographic map of appropriate scales and a corresponding circle sheet, a simple and practical method called the "circle interception" can be used. The method is explained in section 7.1.

Using air photos and a stereoscope and with a slope scale model or a parallel wedge it would also be possible obtain slope values, but this requires a highly trained photogrammetrist for satisfactory results.

Climate, hydrology and water resource data

Climatic data such as precipitation, wind, evaporation, temperature, humidity, etc. can normally be obtained from climatic stations in the watershed or nearby. For watershed management, especially for run-off estimation and erosion control, rainfall intensities are required. However, they are often lacking in the upland or mountain watersheds. In this case, some supplementary automatic rain gauges may need to be placed in the watershed area.

Unless there is a big engineering project under way, data on streamflows and sedimentation are often not available in upstream watersheds in developing countries. Many times, the investigator can only collect data from stations at lower reaches of the same stream or from neighbouring watersheds.

For information on flood damage, drought and other hydrologic problems, the usual techniques are visiting damaged areas, interviewing people and discussing the matter with knowledgeable institutions in the area. The water balance or water budget of the watershed should be estimated. Water use problems regarding quality or quantity need also to be collected.

Land use, land capability and biological data

This category of data usually includes present land use, land use history and future trends, land capability or suitability and a number of vegetation surveys. For present land use, a new survey is often required in order to identify forest and range cover, cropland, plantations, recreation and wildlife reservations, urban and water areas, etc. The needed data and survey criteria for each major land use will depend on management objectives and individual conditions.

Land use history is needed to reveal past lessons and experience. This kind of data can be obtained from reports, records, or from knowledgeable persons in the local community. Future trends in land use are very important to planners. Trend prediction requires estimating population growth, forecasting migration and development, and surveying farmers' intentions for changes.

Land capability or suitability data are usually required to show the limit, extent, and proper use of each piece of land in the watershed. Although criteria may vary, they are essentially based on soils, slope, land capability classification which has been used in many hill watersheds in developing countries is given in Example 4 in 7.1. Appendix 2 shows the classification scheme in summary form. Also in section 7.1, an explanation of land suitability classification framework is given.

Special biological, vegetation and multiple use surveys are sometimes needed. They may cover forest resources, protection forest, range and grassland, wildlife reservations and recreation areas. These surveys are discussed in more detail in section 7.3.

Erosion data

Since erosion is a major concern in most watersheds, the collection of erosion data becomes a very important part of the overall surveys. The causes of erosion should first be identified. They may include many activities of human beings such as cultivating, grazing, logging, mining, road building, housing, fire, recreation activities. Nature also causes erosion in the watershed in the form of landslides, stream cutting, wild-fire, etc. Some of the erosion may be caused by a combination of man and nature. In Chapter 7, survey techniques are introduced and examples provided regarding the collection of data on sheet and gully erosion, road erosion, landslides, stream erosion, torrents, etc. Survey of geology and land forms or geomorphology, both of which are related to erosion, are also explained, in Chapter 7.

4.4 Analysis of major biophysical problems

Land use vs land capability

Determination of proper land use based on land capability or suitability is always the first step toward the protection and development of a watershed. A land use adjustment map can be produced by superimposition of land use and capability maps. Land showing serious over-use should receive urgent attention. On the other hand, land which is presently under-used can be used more intensively. In case of public lands, those under-used can be designated for resettlement of farmers who are cultivating steep slopes or encroaching upon forest lands. The map will not only show the sites, extent and seriousness of the problem areas, but will also provide the basis for rationalization of use of watershed lands. Land presently being used within capability but needing soil conservation treatments will also be shown on the map, and can be used for planning soil conservation activities. Details of such survey and planning can be seen in sections 7.1 and 8.3.

Water resources and use problems

From the basic data collected, an analysis should be made of streamflows including annual and seasonal, maximum and minimum, and qualities such as turbidity, types and sources of pollutants, etc. The timing and frequency of flood and drought should also be studied. Any water use problems, including questions regarding rates of use and problems of quantity and quality, should also be addressed. Section 7.3 shows some examples of analysis. For more details, a water resources and a hydrology book should be consulted.

Hazards of erosion and sedimentation

The various sources and damages of erosion and sedimentation should be identified and analysed, and potential hazards should be pointed out. The latter is very important since most watershed rehabilitation or protection is centred on the minimization of potential hazards. Special efforts should be made to analyse the data collected on erosion and sedimentation. A general methodology may include the following:

- compiling data from the field surveys, observations or from interviewing people;

- analysing soil loss and run-off plot data in the area;

- using erosion models or soil loss prediction equations to estimate quantities;

- analysing storm frequencies, sediment delivery ratios and yields, etc., from the existing hydro-meteorological data;

- compiling reservoir or water storage sedimentation data;

- using geology and geomorphology data for estimating landslide hazards;

- estimating results from all the above data.

The cost of erosion and sedimentation, treatment needs and the benefit of minimizing or controlling them should eventually be estimated.

4.5 Results and products

Maps, statistics and interpretations

Maps and statistics are a major product of this type of biophysical survey. The kinds of maps and statistics produced depend on watersheds and management objectives. The fundamental ones may include the following:

- *base map*, showing boundary, sub-watersheds, villages, roads, etc;

- *topographic map*, showing contours, elevations, land forms, streams, etc;

- *soil map*, showing soil types and boundaries, depths and soil limiting properties;

- *climatic map*, showing mainly rainfall, but statistics may include temperature, evapotranspiration, etc.;

- *geology map*, showing rock types, structures, displacement, morphology, etc.;

- *slope map*, showing different slope classes or exposures/aspects;

- *present land use map*, showing major land uses and cover types;

- *land capability or land suitability map*, showing different land capability classes; or land suitability classes:

- *land use adjustment map*, showing land being over-used or under-used and adjustment needs;

- *erosion or sediment source maps*, showing sites of various types of erosion and sediment potential areas;

- *hydro-meteorological network map*, showing the location of climatic and stream gauging stations;

- *water resource map*, showing surface and underground sources.

Many other maps and statistics of a detailed nature can also be added according to the needs. Some examples are forest and vegetation; landslide potentials; slope stability; stream profiles; land ownership; demographic and population distribution. Many maps can also be combined together.

To produce these maps and statistics on schedule, a flow chart is sometimes needed to set the sequence for collecting data and developing the needed maps.

Brief and essential interpretations are needed for the data collected and analysed. The interpretations should be relevant to watershed problems or management objectives.

Plans and management recommendations

Based on survey data and the results of analysis, the team or teams responsible for biophysical surveys should make draft plans (including treatment, costs, etc.) and management recommendations for further discussion. Planning approaches, techniques and management recommendation details will be discussed in Chapters 8 to 11.

5. SOCIO-ECONOMIC AND INFRASTRUCTURAL DATA

5.1 General

Review of existing reports

Existing reports on general socio-economic conditions of the rural areas of the country should be collected and reviewed before beginning detailed studies in a specific watershed. The existing reports will give managers or planners much basic information which may be valuable to the preparation of survey proposals, related forms and questionnaires.

If there are specific socio-economic and infrastructure reports covering the selected watershed, these reports will be particularly useful. Farmers are usually tired of being asked similar questions over and over again and changes in rural areas are often slow. Therefore, planners should make good use of what information already exists and keep new surveys, especially those of a general nature, to a minimum.

Survey methodology and techniques

Sampling techniques for socio-economic and infrastructure surveys are similar to those discussed in the previous section on biophysical surveys, i.e. random, stratified and cluster. Some socio-economic survey techniques can be seen from FAO Conservation Guide 8. Techniques for infrastructure surveys are illustrated in Section 7.4.

Design and use of questionnaires

Questionnaires need careful design and clear thinking. They should be concise and constructed in a logical order. Ambiguous questions should be avoided. For example, asking farmers whether they need cash subsidies to accomplish tasks they never knew will not only raise false hopes but also get irrelevant answers. Questionnaires should include a double checking system so that if the enumerator detects an inconsistency in the responses, the uncertainty can be resolved immediately. It is often difficult to obtain economic figures, especially those regarding the income of the farmer. To win the trust of the farmer may overcome such difficulties.

5.2 Collection of data

Social conditions

The following are some important social data, among others, that the managers or planners may need to find out in order to draw up a useful plan:

- what will be the population trend in the watershed, its rate of growth, age structure, migration possibilities and other demographic factors that will affect the rate of resource use?

- what are the possible barriers toward innovative technology: poverty, lack of education, poor extension services, tradition, non-aggressiveness, lack of encouragement and incentives?

- what social factors constrain the development and management of the farms in the watershed - land tenure, government rules, traditional farming systems, fear of risk, or others?

- what do the existing social structures, systems or hierarchy influence the individual or community development in the watershed?

- what do the farmers see as their immediate needs - more roads, domestic/irrigation water, housing, marketing arrangements, recreation facilities? ,

- do farmers like to work together or tend to be individualists?

- what is the status of women in the society and their responsibilities? What are the conditions of youth including rates of unemployment, willingness to undertake field work and migration trends?

- to what extent are the farmers aware of the causes and problems facing the watershed?

- what are the farmers' views on the protection and development of the watershed as a whole?

Economic status

Collection of baseline economic data can, in many cases, be combined with the sociological survey. In fact, many social and economic data are interlocked and difficult to separate. The main topics to be covered in a survey of the economic status of a watershed include but are not limited to the following:

- the present economic activities in the watershed, including farm production, farm income, farm models, farming systems, land use patterns, employment, labour demand and supply, rural enterprises, marketing, etc.;

- the potential for economic improvement or development, including farmers' capabilities (labour, resources and technology), non-farm employment opportunities, infrastructure needs, availability of credit or financial aid and agro-industrial development possibilities;

- the constraints or problems of development from an economic point of view, including land tenure, land rental, farm size and fragmentation, capital, knowledge, labour, prices, markets and transportation, etc.;

- farmers' reaction to proposed economic improvement measures including credit and/or subsidies, extension services, taxation and rental reductions, farming equipment and materials, better marketing arrangements, etc.;

- various costs of cropping and farming activities and their returns, the cost and benefit of watershed conservation work, and other related economic figures.

Chapter 7 includes a brief description of socio-economic baseline surveys, farming systems and community development survey, and some other surveys and examples.

Infrastructure in the watershed

The existing infrastructure in a watershed needs to be surveyed. For watershed projects involving rural development, detailed investigations of roads, housing and water and energy supplies are often needed. Some examples are given in Section 7.4.

5.3 Analysis of problems

Survey data should be used to analyse major problems and their possible solutions. Merely presenting data may attract academic interest but it is not good enough for practical watershed management.

Problems needing long-term solutions

Special attention should be given to those socio-economic problems which need long-term solutions, including:

- land tenure. Usually, farmers who do not own the land are reluctant to adopt any soil conservation or protection practices. However, in case of squatting public land, permission should be granted for leasing cultivable lands after classification, provided the farmer agrees to apply and maintain prescribed conservation measures on a continuing basis. This has been done in many countries to end the deadlock of squatting. On the other hand, farmers who presently farm public land not suitable for permanent cultivation should receive first priority for resettlement. In the case of privately rented land, the land owners and the tenants should both be involved in the planning process;

- farming systems of the watershed should be studied and analysed to see whether they are compatible with the principles of sound watershed management. New systems may be developed to benefit both farmers and the watershed. This may require on-farm trials and demonstrations and will take several years to implement;

- farmers' attitudes toward new techniques, extension personnel and government schemes are important subjects for consideration. Many farmers are skeptical about government schemes based on negative experiences in the past. Without the full and willing participation of the farmers, any project will eventually fail. To change farmers' attitudes or to win their trust is a long and gradual process. Possible strategies such as farmer education programmes, better and more active extension services, incentives, and removal of social obstacles should be considered;

- in heavily populated hilly watersheds, a great portion of the land area may already be misused. To correct this requires a long-term approach. Usually, it is beyond the ability of the government to move large numbers of farmers out of the watershed without creating social disorder. Many countries may simply not have suitable land to resettle them. The solution is gradual land use adjustment starting with some simple, scientific and down-to-earth criteria. Technical assistance and incentives should also be planned for such task;

Depending on the actual situation, the planners will need to address many such problems clearly and seek possible solutions. In a heavily populated watershed, population education or family planning will also be a part of a long-term solution.

Problems which can be solved in a relatively short time period

There are many problems which can be solved in a relatively short time period, including:

- problems of infrastructures including roads, water supplies, market facilities, etc., can be alleviated or improved by public investment;

- problems of availability of farming tools and materials - seedlings, fertilizer, pesticide, improved ploughs - can be pre-arranged or secured before a project becomes fully operational;

- capital problems of the farmers. The types of loans or credit, mortgage needs, interest rates, sources and period of loans and repayment schedules, etc., need to be studied fully before making plans and recommendations;

- in addition to credit, other incentive requirements such as subsidies for adopting new practices, tax exemption for farms applying conservation measures, etc., should be considered in order to encourage farmers' participation;

- in many developing countries, unemployment and labour shortages are a major problem in upland watershed areas. A careful analysis should be made to see what can be done to alleviate the problem. Possibilities may include adopting a group approach, public employment to do conservation work on both private and government land, organizing cooperatives, employing small farm machinery, etc.

Expected accomplishments

After a close and joint examination of the above mentioned problems, the managers and planners should make some realistic estimates as to what can be accomplished during the life of the proposed project. This is one of the most difficult tasks the planners must face. Predictions which are either too optimistic or too pessimistic can only hurt the proposed project.

If farmers' acceptance is a major issue, for instance, the planner should not only emphasize education and extension at the beginning of the project, but also keep the goals of the initial period reasonably low. Another strategy is to set a basic annual goal for the first few years and review it every year for modification according to the progress and output of the previous year.

5.4 Result reporting and recommendations

Result reporting

The results of surveys, analyses and findings should be periodically reported to the steering committee for discussion and study. This kind of information is often very useful for other teams in drawing up their respective plans. Any serious socio-economic and infrastructure problems should be brought up early in the planning stage. In many cases, they are much more complicated than technical problems and need policy support from the government. Socio-economic and infrastructure information will constitute an important part of the interim report as well as of the final plan.

Management recommendations

The report on socio-economic findings should include management recommendations. Infrastructure needs should also be addressed clearly.

Future study needs

Periodic socio-economic surveys will be required for monitoring purpose and for detecting changes over time and the impact of the project. Therefore, recommendations on time of future surveys, methodology and criteria based on the first survey should also be spelled out in the team report.

6. INSTITUTIONAL AND CULTURAL INFORMATION

6.1 Information on institutions and legislation

General

Any serious problem regarding institutions and culture should be well studied and analysed. Generally speaking, cultural problems can only be solved over a long time period. In many cases, a project may need to go along with rather than against cultural barriers at the initial stages and gradually influence or change them later.

Institutional problems, on the other hand, can often be solved in a relatively short time period if the government has the political will and firm commitment. The main institutional problems in developing countries often include the following:

- inadequate support from the higher authorities in terms of policy, funding and administration;

- insufficient numbers of trained personnel to carry out planning, design, implementation, field supervision, monitoring and evaluation;

- weak planning and appraisal activities resulting in waste and ineffectiveness in many areas;

- lack of incentives for technicians working in the field;

- poor coordination among related organizations;

- weak field operations due to a lack of efficiency in supervision, reporting and monitoring;

- poor mobility due to lack of vehicles and public transportation;

- lack of research data for continuous improvement.

The above list can be greatly expanded. During planning, such problems should be pinpointed and possible solutions suggested.

Watershed policy and legislation

If there is a well defined government policy on watershed management, it is worth careful study. If such a policy is yet to be formulated, planners should either collect policy statements from related fields such as forestry, agriculture, conservation and water resources, etc., or ask the appropriate authorities to give policy guidance.

Any legislation in watershed management or soil conservation should receive close review. Since watershed management involves resources of many kinds, the planners and managers should review other related legislation as well, including:

- agricultural and related land use laws and acts;

- forest (and rangeland) laws or acts;

- legislation concerning water resource development and use;

- legislation on mining activities and control;

- environmental protection laws and acts;

- recreation and wildlife legislation;

- other related legislation, e.g. on rural development, roads, marketing, etc.

Attention should be given to major conflicts of these legislations toward watershed conservation. In some countries, it may be desirable to propose new legislation or slightly revise existing ones in order to ensure a better institutional set up, or to improve coordination and implementation. For example, in some developing countries, special legislation coordinating forestry, water resources, agricultural and conservations in overall watershed activities has proven very effective. This kind of legislation which changes no basic laws but bridges institutional gaps will usually meet with prompt approval by government authorities.

Legislation regarding the allocation of funds from various sources for watershed conservation use will also be very helpful (see 10.2).

Related government organizations and coordination

The structure and functions of related organizations in watershed management should be analysed. Special attention should be given to those with similar functions or overlapping responsibilities. Any weak organization should be identified by the survey and suggestions for improvement be brought to the attention of the appropriate authorities.

Coordination among related organizations is an important factor and should be subjected to careful study. This should include: history, forms and mechanisms of coordination; successes and failures in the past, etc. From this study a better coordination strategy can be devised.

Present and past programmes in watershed management

This is also a very important subject. Much can be learned from the problems and experiences of present or past projects. Errors must not be repeated. Successful projects merit special attention. Their backgrounds, administration, support services, and field operations need to be analysed.

Education, training and extension needs

Since much watershed work relies on local people for implementation, the importance of farmers' education, training and active participation cannot be over-emphasized. Therefore, the managers or planners should first look at the farmers' needs. For instance, how many extension agents are needed in the watershed to do an effective job? How many farm leaders or contact farmers should be identified and trained, and for what subjects? When should an intensive training programme be initiated and for how long? What kinds of demonstration plots should be established and where and how many?

The planners should also examine the existing quantity and quality of the officers both in specialized agencies and extension service. If their services need to be improved, how many senior and junior officers should be hired? And what type of skill training is needed for both new and old officers?

Sometimes mechanisms and techniques in extension may need to be entirely overhauled in order to deal effectively with the large numbers of small farmers living mostly at less accessible areas. Small farmers will not be easily convinced or motivated by meetings and slide shows alone. Most land conservation measures, for example, require technical assistance and close supervision. Therefore, a thorough review of the existing extension set up, including capabilities, strategies and problems, is absolutely necessary.

Another important issue is the division of labour between specialized field officers and general agricultural extension agents. The extension officers should be geared to do their normal extension duties while technical work is handled by these specially trained professionals or sub-professionals. Fig. 5 shows an example of coordination or division of labour between extension and soil conservation agencies in pursuing conservation farming activities.

Fig. 5

AN EXAMPLE OF COORDINATION OR DIVISION OF LABOUR BETWEEN EXTENSION AND SOIL CONSERVATION AGENCIES

For Conservation Farming Work

(1) Overall planning & budgeting	Joint efforts
(2) Farmer's meeting & Extension	Joint efforts
(3) Farmer's interviewing & farm planning	Extension officers supported by soil conservation staff
(4) Surveying & staking	Soil conservation staff
(5) Supervision construction	Soil conservation staff
(6) Mapping & recording	Soil conservation staff
(7) Soil conservation subsidy distribution	Joint efforts
(8) Follow-up activities (i.e. cropping, credit and marketing)	Extension officers
(9) Maintenance inspection	Soil conservation staff
(10) Farmer's training in soil conservation	Joint efforts

Source: Soil Conservation Division, Jamaica

Local community development

Local communities are important to any watershed project. Assisting in developing local communities will, in turn, benefit the people as well as the watershed. Community development work is always a worthwhile investment in a watershed management project.

Many community development or improvement efforts are relatively inexpensive, yet can mean much to the local community and the well-being of the inhabitants. From the survey of farmers' needs, the planners should be able to work out realistic development plans to cater for community needs within the scope and resources of the project.

Farmers' organizations, community and special interest groups

Farmers' organizations such as farmers' associations, irrigation districts and various bodies which are vital to watershed activities should not be overlooked by the planners. Many farmers' organizations may have their own extension staff and funds to assist various activities in the watershed. Even if they do not have available funds, at least they can represent the view of farmers in the area. Care, however, should be taken when approaching those organizations having narrow interests and strong political views. Involvement of one of these groups may cause the non-cooperation of others.

Community-level organizations such as county and village councils are essential in many watershed activities. They should be contacted and involved in planning and implementation as appropriate.

Special interest groups such as environmental protection or conservation societies, 4-H clubs, women's groups or youth clubs and private aid agencies and non-governmental organizations (NGO) in the area should be contacted and consulted. Some of these groups should be involved in spreading messages, disseminating technical information and supplementing extension activities.

Marketing and others

Some watershed conservation projects may not aim directly at production increases. But, farmers will not be convinced to adopt new conservation measures unless these will result in increasing crop production and income. Otherwise, they will not even maintain existing measures.

Producing more and better crops without appropriate markets is also futile. Even subsistence farmers need cash to buy farm tools and to pay their bills. Yet marketing is often neglected in many watershed conservation projects. The marketing arrangement is mainly an institutional arrangement. It includes crop outlets to factories, agro-industries, purchasing stations, etc. It may also include price support, transport and storage arrangements. All these information should be collected and analysed.

Other institutional information such as land registration, cadastral surveys, and institutions for credit and mortgage arrangements, seedling and fertilizer supplies, etc., are also needed.

6.2 Cultural information and considerations

Compatible with cultural patterns

To consider local culture in planning is to minimize possible resistance in future implementation. Farmers are relatively conservative. Any improvement which is compatible with the local culture and with a gradual path will have better potential for success. For instance, bench terracing in upland areas is easily accepted in the Far East, where rice paddies are common. Planners must carefully collect and study cultural information before making any recommendations for drastic change. A mild and slow change at the beginning, followed with continuing extension efforts is often the best path to pursue.

Traditional practices

Traditional practices have their roots in culture and society. Slash and burn shifting cultivation, for example, has been widely practised in many parts of the developing world. Unless the farmers are provided with alternatives, i.e. lands, farm inputs and technical know-how, this practice will not easily be changed just by passing a law. The subject needs a profound study before alternatives are suggested.

Other traditional practices such as uncontrolled grazing, use of fire to clear fields, and up-and-down tillage may have their reasons. Any substitution or improvement should stem from study or research and must be acceptable and beneficial both to those who apply it and to the environment.

Religious influences

Religious beliefs may affect the behaviour and daily life of the local people in rural watersheds. Churches or temples in many countries exert great influence upon local communities. The best policy may lie in close collaboration with religious organizations. In Thailand, for instance, many educational campaigns and development activities are being done through Buddhist temples.

Cultural activities

Besides economic activities, the information of the people's cultural activities should also be collected and analysed. Sports, music, movies, reading, gambling or other leisure activities may have implications for the planned watershed effort. Sometimes, a watershed or rural development project needs to provide a recreation centre or sports events to pursue better public relations and communication, especially for youth.

Urban and rural relationships

For watersheds in developing countries, urban and rural relationships are usually equivalent to downstream and upstream relationships. Because of the wide cultural gaps, these relationships often present difficulties. For example, youth in rural areas are willing to take almost any job in towns or cities, whereas urban youth, even when unemployed, seldom can be attracted by farm work. However, the physical relationship between upstream and downstream areas of a watershed cannot be separated. Information and consideration should be given to methods for establishing closer links between the populations of the two areas.

Attitudes toward group action

Many upland watersheds suffer labour shortages on a permanent or seasonal basis. Group action has the following advantages:

- many watershed management activities, such as gully control, streambank protection, reforestation and community pasture improvement, etc., can be better carried out by group actions;

- on private farms, group action facilitates mutual drainage systems, better terracing work, especially for broad-based terraces on gently slopes, and farm roads where farm boundaries can be used as sites to serve two or more farms;

- with group action in a sub-watershed, planning, design, implementation and supervision of activities can be much easier, more concentrated, timesaving and effective.

However, the attitudes toward group action may differ from one place to another. During survey, this subject should be carefully studied.

6.3 Analysis and reporting

Organization for effective implementation

After studying various institutional settings and their problems, suggestions should be made as to which organizations should be involved in implementation and how the work can be carred out effectively:

- not all the agencies which participate in the planning of a multi-disciplinary watershed project should be designated as implementing agencies. Only those with major responsibilities should co-sponsor the project. Others may remain as cooperating agencies;

- institutional strengthening should be spelled out for the major responsible organization(s);

- coordination mechanisms should be stipulated in details;

- an evaluation body should be established and should include the steering committee or the liaison unit mentioned previously, and representatives of the local communities.

Suggestions on reasonable course of action

The report of this section should also include suggestions on a reasonable course of action after studying and analysing various cultural aspects mentioned previously. These may include the following subjects:

- a long-term strategy to overcome cultural difficulties;

- a pilot or a test area to obtain local reactions;

- a gradually expanded project;

- an intensive education program.

7. SOME SURVEY TECHNIQUES AND EXAMPLES

As stated in the Foreword, the watershed condition in each country is unique and it is impossible to provide detailed survey techniques which are applicable to all cases. This chapter, however, is intended to introduce some of the basic ones and give examples for reference and for stimulation purposes only. Not all of them are required in one project.

The following five sections present more than thirty survey techniques, grouped under twenty-five sub-headings. In addition, a total of twenty-four examples is given in this manual. Most of them are drawn directly from field experience.

The five sections cover:

- land use, land capability and suitability surveys;

- erosion surveys;

- water and other natural resources surveys;

- infrastructure surveys;

- socio-economic surveys.

7.1 Land use, land capability and suitability surveys

Land use, land capability and suitability surveys are fundamental for rationalizing land use in a watershed. Following are descriptions of five basic surveys:

- soil survey;

- slope analysis;

- land capability classification and land suitability classification;

- present land use survey;

- land use adjustment survey.

1) Soil survey

Soil survey needs will have to depend on whether or not there are existing data available. If a standard soil survey has already been undertaken, only some supplemental data will be required. If no such survey has ever been undertaken, there will be a need for a new survey.

New survey

When soil information is lacking a new survey must be undertaken. For watershed conservation, the main objective of survey is to provide basic information for land capability classification, and therefore, the survey may be of a relatively simple nature. Such a survey usually includes: identification of major soil types and their boundaries; recognition of problem soils; information on soil depth; and identification and location of limiting factors, e.g. stoniness, waterlogging, occasional flooding, severe erosion, etc.

If the survey is for land suitability classification, further information on soil nutrients, moisture and root zone conditions, as well as on economics, etc. should be collected. Soil survey manuals and guideline books of FAO should be consulted.

Example 1 provides simple guidelines for a soil survey of a watershed in northern Thailand. The work was done by university students under a government project supported by FAO. The field survey, covering a total area of 42 700 ha was completed in 150 man-days. A final report and a 1:15 000 scale map were produced soon afterwards. The objective of the survey is for land capability classification using the "Treatment-oriented system" (see Appendix 2).

Supplemental survey

Even if a standard soil survey exists, a supplemental survey may still be needed. For instance, soil depth and soil limiting factors are vital to land capability determination yet they may not be shown sufficiently on the existing maps due either to the scale of the map or the nature of the original survey. Also, soil boundaries will need to be re-checked because standard soil surveys usually concentrate on agricultural lands rather than on watershed slopes.

Example 2 describes the working procedures for a supplemental soil survey implemented in a FAO supported watershed project in Jamaica.

2) Slope analysis

Slope analysis is an important step towards rationalization of land use in a watershed. It provides the basis for land capability classification, land use planning and soil conservation needs. The "Circle Interception Method" which can be done by office clerks with little training is explained in Example 3. A circle is used because it has equal distance in all directions and hence represents a fixed horizontal distance on the ground. By overlaying a contiguous circle sheet on a topographic map and counting the contour intervals in each circle, one can get the values of the vertical rise.

The slope of a circle can thus be determined by the following equation:

Slope (%) = Vertical rise / Horizontal distance

The diameter of the circle depends on the map scale and contour interval. Too big a circle will represent too great a distance on the ground whereas a circle which is too small will make the work impractical. Diameters of 8 mm to 11 mm have been used satisfactorily on 1:10 000 and 1:15 000 scale maps respectively.

Slopes can be grouped into many categories. As a rule, slope categories should be the same ones used in the land capability classification. Each slope category should be assigned a number or a letter and given a colour for mapping convenience. Example 3 shows the techniques for slope analysis and mapping.

Example 1

SIMPLE GUIDELINES FOR NEW SOIL SURVEY

Technical Guidelines of Soil Survey in the Project Area

1. The purpose of this brief soil survey of the project area (approx. 42 000 ha) is mainly for land capability classification.

2. Major soil groups of the project area should be identified and the boundaries of each marked on the map. To find their boundaries the following procedures can be used:

1) Using air photos to find out the possible boundaries indicated by vegetation differences.

2) Travelling on major roads and main foot paths to check the boundaries.

3. The following soil limiting factors should be looked at closely in addition to the major soil types:

1) Soil depths: four classes (>90 cm, 50-90 cm, 20-50 cm and <20 cm) to be measured by a soil auger and marked on the map. If the soils are deeper than 90 cm, no actual depths should be measured and it only needs to mark on the map: D(>90 cm).

Soil depths can also be observed along road cuts, stream banks and housing sites. Usually, for the same type of soil, depth measurements should be made at the changing of topography i.e. ridge tops, side slopes, foothills, etc. No soil depth and other soil data are needed for the Red/Yellow Podzolic soils which are considered to be unsuitable for cultivation because they are severely erosive when so used. Neither soil depths nor other soil data are needed on alluvial soils.

2) Severe erosion: only gully erosion will be noted and the areas be marked on the map.

3) Stony: when the presence of gravels or stones is an obstacle to normal cultivation and soil conservation treatment, the site should be marked on the map.

4) Wetness: including potential flooded areas.

4. Forms and labels:

A simple recording form should be made according to this guideline and to be used in the field. Some reference information may be included in the form, for example, site (grid reference), vegetation, topography, etc.

The sequences of labels on the soil maps are suggested as follows in each mapping unit:

Soil Type - Depth - Other Soil Limiting Factor

For example:

L - D - 0 means:

(Reddish-brown No

lateritic soil - 90 cm - limiting factor)

L - S - S means:

(Reddish-brown

lateritic soil - 20-50 cm - stoney)

A

(for alluvial soils): no other labels needed

P

(for red-yellow podzolic soils): no other labels needed

5. Final products:

1) A map with soil boundaries, soil depths and limiting factors (scale 1:15 000).

2) An illustrative report.

Soils survey information form

Code No.:

Location: (including grid reference)

Soil types: Other types:

reddish-brown lateritic soil

red-yellow podzolic soil

alluvial soil

Soil depth:

D > 90 cm

MD 50-90 cm

S 25-50 cm

Vs < 20 cm

Limiting factors:

Stoney

Wetness (or occasional flooding)

Severe gully erosion

Other information:

Parent material:

Vegetation type:

Topography:

Others:

When soil is less than 90 cm, use auger to measure and mark soil depth class on this sheet and the actual depth on map.

Date: Surveyor:

Example 2

PROCEDURES FOR SUPPLEMENTAL SOIL SURVEY

1. Introduction

Jamaica has completed a Soil and Land Use Survey covering the whole island. The reports were issued intermittently by the Soils Department of the Regional Research Centre, University of the West Indies, Trinidad, on an irregular basis since 1958. The soil classification system used in these surveys was rather simple and conventional, compared with the international soil classification systems now existing. Its advantage was that it was based upon almost entirely physical characteristics (texture, structure, colour, depth, drainage, moisture retention, etc.) of the soils which can be easily recognized in the field.

The present purpose for producing a soil map is mainly for land capability classification. The Project feels that the existing information can be largely used. Some supplementary items such as soil depths, land use limiting factors will have to be collected in the field. Therefore, taking account of the time constraint, it was decided to use the existing maps as a basis for preparation of a new one.

2. Preparation

Before going out to the field, all available reference material and aerial photographs were gathered and studied. Sometimes, a general observation in the field was required. The procedures were:

1) Soil boundaries and other information were directly transferred from annotated photographs to the existing maps of various scale. However, the maps attached to the Parish survey report at a scale of 1:50 000 are the basic material for producing soil maps covering the project area.

2) The existing soil maps were studied and placed over a topographical map in order to distinguish possible inaccuracies in soil boundaries. The boundaries suspected of being inaccurate were also checked using aerial photographs provided by the

3) Limestone and river wash areas can easily be distinguished by the photographs. It was found that, in general, there was estimation of the river wash and an underestimation of the limestone areas. These should be corrected.

3. Field Checking

1) Some soil boundaries were corrected by checking in the field.

2) Soil depths were taken throughout the watershed in order to obtain a general idea of the average soil. depth of each specific soil type in the watershed. On slopes steeper than 30 °, no soil depths are measured.

3) Finally, those areas with severe soil limiting factors preventing intensive land use were indicated on the maps. Soil limiting factors taken into consideration where stoniness, occasional flooding, waterlogging and severe erosion hazards.

4. Final Mapping and Statistics

The final soil map was drawn at a scale of 1:10 000. Since the final soil map was allowed to produce in a period of two to three weeks for each unit watershed, it was impossible and undesirable to have great precision and detail. It is important to note that the accuracy of the prepared soil map will not exceed that of the existing map drawn at a scale of 1:50 000 especially regarding soil types and their boundaries.

The most important elements for this type of map, however, are soil depths and limiting factors. Finally, statistics were prepared, giving the acreage covered by each soil type or mixture of soil types in the area. This was done.

by cutting out the different areas of each soil type and weighing them against a reference square of known acreage.

Example 3

SLOPE ANALYSIS AND MAPPING

1. Introduction

The main objective of slope analysis is to provide the basis for land capability classification and for planning of proper land use and soil conservation treatment.

The importance of slope analysis cannot be over-emphasized since this is the first step towards rationalized use of watershed slopes.

Many methods are applicable for such analysis: airphotos; slope models; even digitized computers. However, judging from resources available and the manpower conditions in the developing countries, the "Circle Interception Method" is probably most suitable.

2. Materials required

1) Topographical map: using 1:10 000 and 25ft (7.5 m) contour topographical map for this example.

2) Circle sheet: a transparent film (or paper) printed with contiguous circles of 0.3 inch (8 mm) diameter as an overlay on the topographical map for analysis.

3) Slope analysis graph: 2 a slope analysis graph is produced in the following way (for the use of 25 ft or 7.5 m contour and 1:10 000 scale map): (See page 51).

a. A linear line is established on a piece of grid paper between the intercepted contour intervals and slopes:

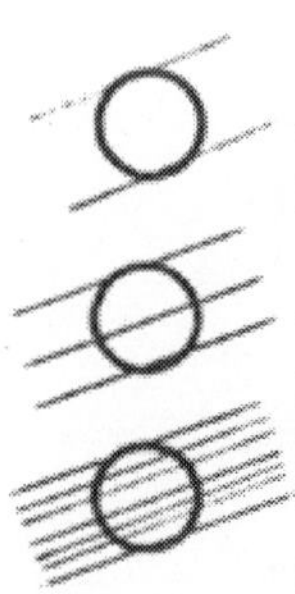

One interval: $\frac{25}{250}$ = 10%

Two intervals: $\frac{50}{250}$ = 20%

↓

Six intervals: $\frac{150}{250}$ = 60%

b. The slope is divided into seven classes and then each line is extended to meet the linear line mentioned above.

c. Using the linear line as an axis, the relation between slope classes and the corresponding contour intervals is obtained.

2 Different scale of map, contour intervals and/or diameters of circle need different graphs. However, the procedure to produce a graph is the same.

3. Slope categories

Slopes are divided into seven categories or classes in line with the land capability classification criteria. Each class is assigned a colour.

1 <7°	Yellow	2 7°-15°	Purple
3 15-20°	Green	4 20-25°	Brown
5 25-30°	Blue	6 30°-40°	Red
7 >40°	Orange		

4. Smallest area

One circle on the 1:10 000 map equals 1.126 acres (0.45 ha). With the surrounding area of 0.182 acres it becomes 1.3 acres (0.52 ha). The smallest area of 1.3 acre (0.52 ha) was thus adopted because the air photo interpretation of present land use has employed the smallest area of 5 mm by 5 mm, or 1.4 to 1.9 acres (0.56 ha to 0.76 ha) on different scales of photos. Since the circle's diameter represents 250 feet (76 m) on the ground, the slope thus obtained was the average of that distance unless in some occasions two slopes were analysed in one circle.

5. Analysis procedures

1) Superimpose the 'circle sheet' on the 1:10 000 scale and 25 ft (7.5 m) contour map. All the sheets should be mounted in a consistent manner in one watershed.

2) Count the contour intervals intercepted by each circle. Special attention should be given if there are two slopes (ridge or valley, etc.) in a circle.

3) Determine the slope category of each circle from the slope analysis graph. First read horizontally according to the contour intervals counted from the circle. When this meets the linear line, read down to get slope category or class.

4) Give each circle a proper colour.

5) Group the same coloured circles on another overlay for producing a draft slope map.

6) Checks have to be done in three stages:

1st stage: Before grouping, a systematic sampling check should be undertaken by a supervisor. A record should be kept to see the accuracy of the work. If low accuracies (less than 90X) are found, the analysis work of that sheet should be re-done.

2nd stage: After the draft map is produced it should be again superimposed on the original topographical map for carefull checking.

3rd stage: A spot check in the field is needed at several dozen points in the watershed. These points should be accessible, easily recognizable and well distributed if possible.

7) Draw a final map with slope classes (1 to 7) labelled for each topographical unit.

6. Progress and efficiency

Any draughtsman or person with little experience can be trained to do this kind of analysis work. According to experience, after several days to one week of training, a person could complete about 400 circles an hour. Based on a five-hour working day, 2 000 circles or an equivalent of 2 600 acres (1 040 ha) on 1:10 000 scale map can be completed each day. A small watershed of 25 000 acres (10 000 ha) can be completed in two weeks time (or in ten working days) with a two-man team.

SLOPE ANALYSIS GRAPH

To be used for Scale. I: 10,000 Contour interval 25 feet. Circle diameter 0•3 inch.

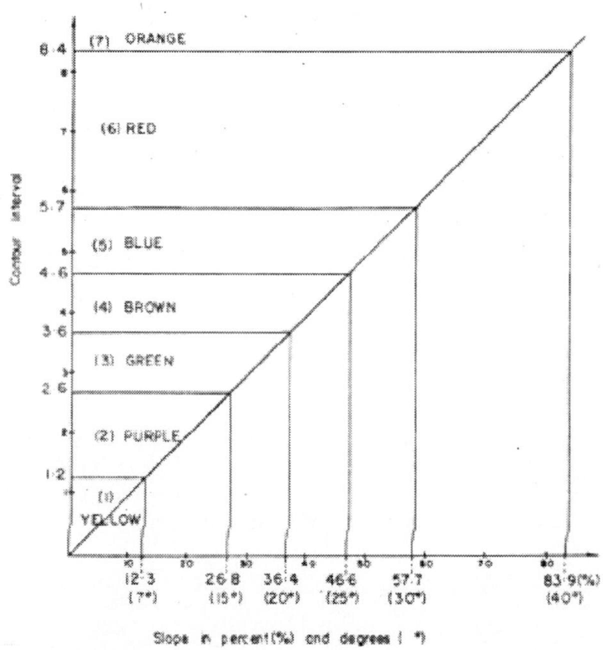

Slope analysis can also be done by using air photographs and stereovision. The photos are first aligned. Specially designed 'slope scale models' are placed on each print and rotated until a sloping dotted line fits the slope to be measured. The angle between the two dotted line is read to determine the actual slope on the ground. The principle employed is that of a variable parallax wedge. It needs an experienced photogrammetrist to do a satisfactory job.

Computer programs for terrain analysis are available when Geographic Information Systems (GIS) are used.

3) Land capability and suitability classification

Land capability and suitability classification are the foundation of proper land use. Many classification criteria have been developed since the first one was introduced in 1930s by the Soil Conservation Service of the United States. Although capability and suitability are sometimes exchangeable, the former's primary consideration is to prevent land degradation and the latter is to consider the fitness of a given type of land for a defined use (FAO, 1976a).

A "Treatment-oriented capability classification"

For hilly watersheds, a "treatment-oriented" classification has been used successfully in many developing countries since the 1970s. The characteristics and usage of this classification can be briefly described as follows:

- The land is classified according to two major factors: slope and soil depth. When a third factor, soil limiting factor, is present, the land is classified as suitable only for less intensive use. All these factors can be measured or seen on the ground and the process and results of classification can be easily understood by field assistants and farmers.

- Each classification is accompanied by land treatment requirements. A piece of land which cannot be treated with prescribed conservation measures should not be used for cultivation or orchards.

- Land is classified to its most intensive use permissible. It is permissible for less intensive use but not for over-use.

- The classification can be quickly learnt and applied by semi-skilled workers to find lands suitable for cultivation, orchards, pastures or forest at the watershed level either for land use adjustment, settlement or development purposes. It can also be applied readily at the farm level by a field assistant using a hand level to measure slopes and a soil auger to examine soil depths.

Appendix 2 shows the classification in a summary form. A detailed report can be seen from Sheng (1972). Example 4 illustrates the procedures for producing a land capability map using this classification system. Figure 6 shows the mapping procedures for producing this type of land capability map and thereafter a land use adjustment map.

Example 4

PROCEDURES FOR PRODUCING A LAND CAPABILITY MAP

1.Introduction

The procedure is for developing a map based on the treatment-oriented land capability classification system for classifying steep upland watershed! under a homogeneous climatic condition.

2. Some basic principles

1) Major factors to be used for the classification are:

a. Slope;

b. Depth of soils;

c. Other limiting factors i.e. stoniness, wetness, gully dissection, frequent flooding, etc.

2) Land which has a slope of less than 25 degrees and deep soil which is treatable with prescribed soil conservation measures is classified as C (cultivable). From 25 degrees to 30 degrees, if it is treatable the land is classified as FT (food/fruit trees). Any land which is over 30 degrees (or below 30 degrees untreatable) should be classified for other purposes i.e. forest, pasture, agroforestry etc., for permanent vegetative cover.

3) Each class of land so classified has implications for:

a. Type(s) of soil conservation needed;

b. Tools to be used for construction, conservation treatments and cultivation;

c. Prospects for mechanical cultivation.

4) Land is classified to its most intensive use. Use within its capability and lower use are permissible but use beyond its capability should be discouraged.

5) Soil conservation treatment specification tables (in FAO Watershed Management Field Manual 13/3) which have depths of cut or minimum soil depths can be used for guidance in classifying land capabilities.

3. Procedures

1) On an illuminated table, put down a coloured slope category map.

2) Place a permatrace soil map (with information on soil

boundaries, depths, other limiting factors, etc.) atop the slope category map.

3) Overlay another blank permatrace or transparent sheet for producing a draft land capability map.

4) Look at every slope category as a unit, consider its soil depths and other soil limiting factors, i.e. stoniness, wetness and erosion from the soil map.

5) Use the check list (on page 56) for classification guidance. If necessary, the "Specification tables" mentioned before can also be used for reference.

6) Determine the proper classes starting from the obvious ones as follows:

   a. Regardless of soil depths and other limiting factors:

      Slope 7 (orange colour) <u>classify to</u> ——→ FP (protection forest)
      Slope 6 (red colour) ——→ F (forest land)

   b. Where soils are deeper than 90 cm (36") and there is no soil limiting factor, the classification is straightforward as follows:

      Slope 5 (blue colour) <u>classify to</u> ——→ FT (for fruit or food trees)
      Slope 4 (brown colour) ——→ C4 (cultivable land 4)
      Slope 3 (green colour) ——→ C3 (cultivable land 3)
      Slope 2 (purple colour) ——→ C2 (cultivable land 2)
      Slope 1 (yellow colour) ——→ C1 (cultivable land 1)

   c. Regardless of soil depth, where soils have limiting factors preventing normal tillage i.e., too stony, too wet, frequent flooding and severely gully dissection, the classification is as follows:

      Slope 1, 2, 3, 4, ——→ P (pasture)
      Slope 5 ——→ F (forest land)

d. Where (1) soils are shallower than 90 cm (36") and (2) without soil limiting factors, the classification will be done as follows:

- Slope 5
  - → AF (soil depth not enough for 1.75 m wide orchard terraces)
  - → FT (enough for above treatment)
- Slope 4
  - → P (soil depth not enough for 2 m wide hillside ditches)
  - → C4 (enough for above minimum treatment)
- Slope 3
  - → P (soil depth not enough for 2 m wide hillside ditches)
  - → C3 (enough for above minimum treatment)
- Slope 2
  - → P (soil depth not enough for 2 m wide hillside ditches)
  - → C2 (enough for above minimum treatment)
- Slope 1
  - → P (soil depth below 15 cm (6"))
  - → C1 (soil depth above 15 cm (6"))

7) Special consideration is given to extreme conditions of soil texture on steep slopes such as too friable or too heavy. In these cases, slope 5 should be classified to AF regardless of depth.

8) Make a final copy of land capability map in black ink.

Note: The minimum treatment and minimum soil depths required can vary according to local needs.

CHECKLIST FOR PRODUCING A LAND CAPABILITY MAP

(At watershed level)

Slope Category	Colour	Soil limiting factors*	Soil depths	Land cap.
7	Orange	classify to		FP
6	Red			F
5	Blue	(a) with factor(s)		F
		(b) without	(1) <Min. depth 70 cm (27.5")	AF
			(2) >Min. depth 70 cm (27.5")	FT
4	Brown	(a) with factor(s)		P
		(b) without	(1) <Min. depth 67 cm (26.4")**	P
			(2) >Min. depth 67 cm (26.4")	C4
3	Green	(a) with factor(s)		P
		(b) without	(1) <Min. depth 56 cm (22")	P
			(2) >Min. depth 56 cm (22")	C3
2	Purple	(a) with factor(s)		P
		(b) without	(1) <Min. depth 47 cm (18.5")	P
			(2) >Min. depth 47 cm (18.5")	C2
1	Yellow	(a) with factor(s)		P
		(b) without	(1) <Min. depth 15 cm (6")	P
			(2) >Min. depth 15 cm (6")	C1

* Such as S (stony), W (wet), G (gully dissected), F (flooding) and O (others).

** Min. depth = $\frac{2\ m}{2} \tan 25° + \frac{0.2\ m}{2} + 0.1\ m$ (for safety) = 0.67 m (26.4")

SYMBOLS FOR LAND CAPABILITY

C1	Cultivable Land 1
C2	Cultivable Land 2
C3	Cultivable Land 3
C4	Cultivable Land 4
P	Land for Pasture
FT	Land for Food Trees or Fruit Trees
AF	Land for Agro-forestry
F	Land for Forest
FP	Land for Protection Forest
U	Urban Land

Note: The map at the right is only a part of Hermitage Watershed, Jamaica

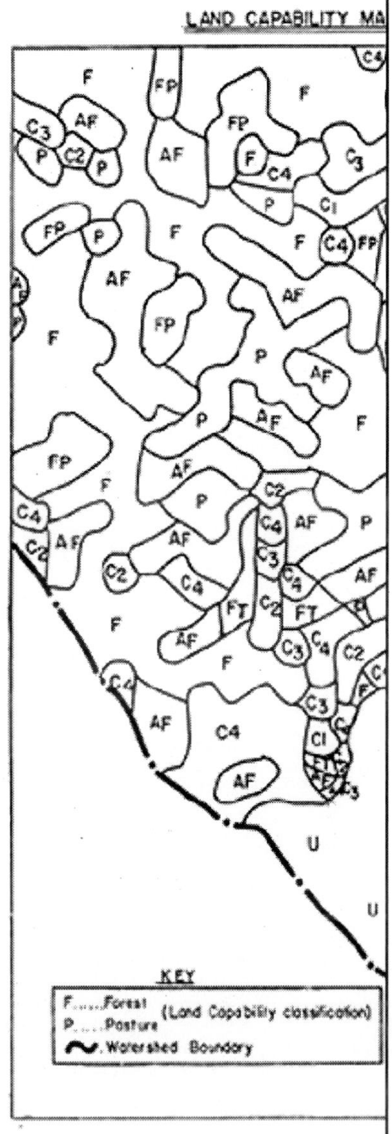

Fig. 6

DIAGRAM OF MAPPING PROCEDURES FOR LAND CAPABILITY AND LAND USE MAPS

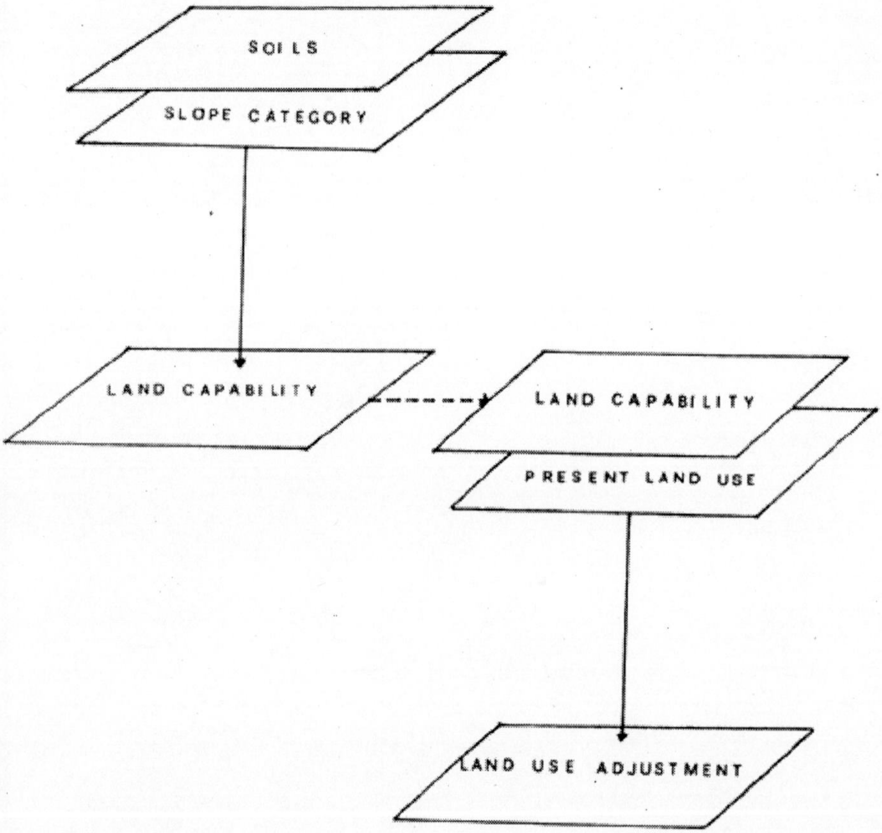

Note: For watersheds under homogeneous climatic conditions. Otherwise, climatic factor should be added to determine land capability and land use.

Land suitability classification

Land suitability classification is the process of appraisal and grouping of a given area for a specific kind of land use (FAO, 1976a). Economic considerations, among others, are strongly involved in the determination of suitability.

There are several levels of classification according to actual needs for finding varying degrees or categories of suitability. The four categories are as follows:

- *suitability order:* Indicating whether land is assessed as suitable or not for the major kinds of land use such as rain-fed agriculture, irrigated agriculture, grassland, forestry or recreation. There are two orders represented in maps and reports: S (suitable) and N (not suitable);

- *suitability class:* Indicating degrees of suitability within order. For instance, S1 (highly suitable), S2 (moderately suitable), S3 (marginally suitable), N (not suitable), etc. To determine the class, economic assessment is needed, i.e. inputs, benefits, and net incomes, etc.;

- *suitability subclasses:* Indicating kinds of limitation such as m (moisture deficiency), e (erosion hazards), n (nutrient problem), etc.;

- *suitability unit:* Indicating subdivisions of a subclass. The units differ from each other in their production characteristics or management requirements.

The criteria used can be qualitative or quantitative. Also, the classification can be applied to the current use or potential use. A piece of land can be classified into many objectives according to the needs. Fig. 7 shows a summarized structure of suitability classification and Example 5 shows several suitability maps of the same piece of land. Details can be seen from FAO Soil Bulletin 32: A Framework for Land Evaluation. Suitability criteria for rainfed agriculture, irrigated agriculture, or forestry can also be seen from FAO publications (1983b, 1985, and 1984b).

Figure 7

STRUCTURE OF THE SUITABILITY CLASSIFICATION

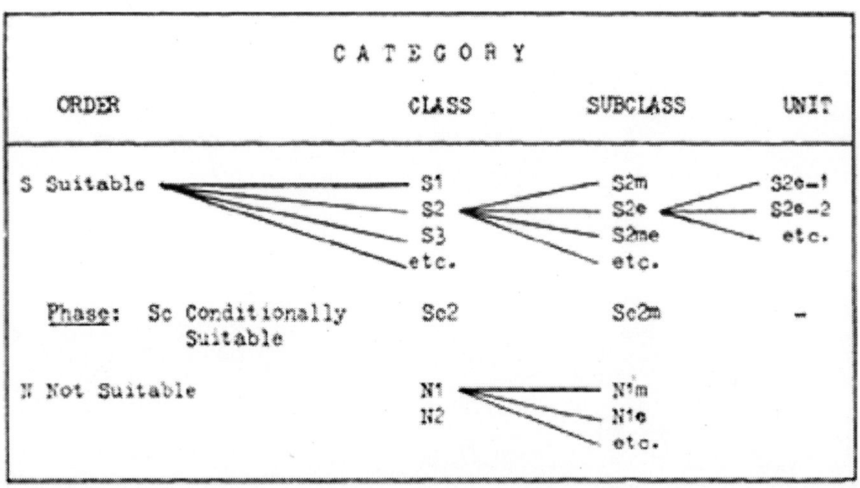

Source: FAO 1976a

Example 5 EXAMPLES OF QUALITATIVE LAND SUITABILITY MAPS

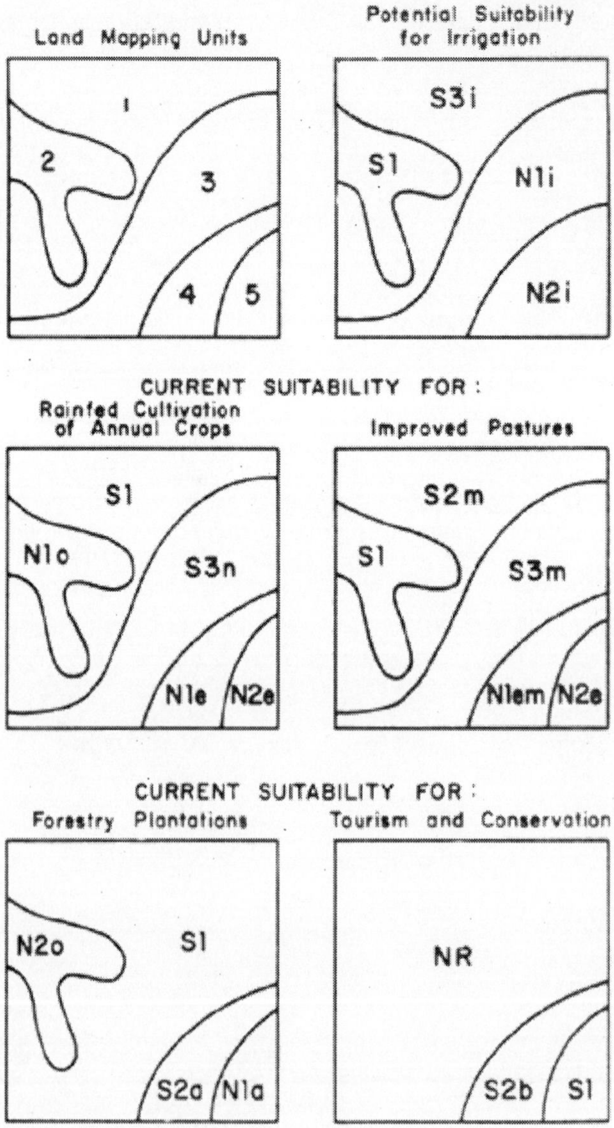

NR : Not Relevant b: Biological Resources

(Source: FAO. 1976a)

4) Present land use survey

Present land use and cover type surveys are also fundamental to watershed management. Before beginning this type of survey, several important factors must be predetermined:

- the scale of the map. It should be the same scale as other basic or related maps in order to permit overlays and/or comparisons;

- the scale of air photos. It should be convenient for image transfer;

- determine major land use and cover types to be surveyed and each should be given a mapping symbol. An example is given with the present land use map (page 66);

- decide smallest area for interpretation and mapping.

During the survey, attention should be given to the following:

- at the beginning, close checks should be made to establish the relationship between photo images and actual ground conditions;

- determine major use type for mixed cropping patterns;

- the difference between fallow lands and grasslands;

- the difference between natural and human disturbed areas.

Example 6 shows some details in obtaining aerial photographs, survey procedures and mapping, under a FAO project in Jamaica.

Using computers with image processing techniques and Geographic Information Systems (GIS) can also produce this kind of map. Explanations can be seen in Appendix 3.

Example 6

USING AERIAL PHOTOGRAPHS FOR SURVEYING PRESENT LAND USE AND MAPPING.

1. Aerial photography

The project contracted a Canadian firm to do aerial photography. A brief description of photography and photos are as follows:

1) Before flying and photo-taking by a Canadian firm, 50 triangulation points were marked on the ground by the project. These points were scattered over the project area and some were outside but close to the watershed boundary.

2) The actual photography was done in December 1980. A total of 15 flights were made.

3) Two sets of photographs were obtained in February 1981, one in black-and-white and the other in natural colour. After selection, the final prints of natural colour were obtained in May 1981 in two kinds of colours. The black-and-white set was for producing contour maps.

4) Each set has a total of 230 photos (for 90 000 acres).

5) The photo scale ranges from 1:15 000 to 1:18 000 averaging about 1:17 500.

6) Overlapping of photographs in the flight line was about 65 percent and 75 percent; between two flight lines was 20 percent to 45 percent and these met with the usual standards and requirements.

2. Photo-interpretation

The photo-interpretation work was done in July and August 1981. The procedures were as follows:

1) The centre area of each odd (or even) photo was marked for minimizing overlapping of work.

2) Transparent films were placed on every right photo of a stereo pair. Delineation of land use patterns was done with a rapidograph pen.

3) Frequent field checking was made to ascertain the land use types. Usually after a day of delineation in the office, the following day was spent

on field checking.

4) Discrepancies between delineation on adjacent air photos, in and between flight lines, were carefully corrected to enable an accurate transfer of information to the photo mosaics.

5) Special remarks on photo-interpretation:

a. The smallest unit delineated on photos was 5 mm by 5 mm or about 1.4 acres to 1.9 acres (0.56 ha to 0.76 ha).

b. Mixed crops which cannot be seen from the air photos created some difficulty in identification and delineation. Therefore, only the apparent mixtures were interpreted.

3. Mapping

The present land use map indicates various types of information, pertaining to land use at time. However, the accuracy and effectiveness of this map, is dependent on several factors, namely:

1) Availability of up-to-date high quality aerial photographs.

2) Good and thorough interpretation of the photographic imagery.

3) Size of the smallest area which can be shown on the map.

4) Type of transfer method i.e. by eye or by instruments such as a Sketch Master.

4. Materials

1) Aerial photographs: 1:15 000 or 1:18 000 scale natural colour photographs.

2) Mosaics: black and white 1:10 000 scale. These are controlled mosaics produced from the same photos mentioned above.

3) Clear (stable) film, drawing pens, etc.

5. Methodology

1) Gather photographs, mosaics of area (watershed) to be worked on.

2) Secure mosaic to drawing table, or another flat work surface.

3) Overlay and tape down a sheet of clear film.

4) Draw on registration corner marks, boundaries, margins, etc.

5) Depending on which section of mosaic is selected, find the corresponding photograph(s).

6) Observe the land use type or class which has been interpreted and boundaries which have been drawn on the photograph.

7) Identify the said type or class and its area on the mosaic.

8) Carefully and accurately draw the corresponding line or boundary(ies) onto the clear film. The line should be thin and lear. It is best to use a rapidograph pen 0.3 or 0.4 in point size.

9) On completion of a land use type or area, write in the proper symbol.

SYMBOLS FOR PRESENT LAND USE

Annual Crops	CA
Perennial Crops	CP (General)
Coffee	Cf
Banana	Ba
Citrus	Ci
Sugar-cane	Su
Coconuts	Co
Fallow/Natural Pasture/Grasslands	G
Improved Pastures	PI
Food Forest	FF
Forest and Woodland	F (General)
Conifer Plantations	Pc
Hardwood Plantations	Ph
Natural Hardwoods	Fn
Bamboo	Bo
Ruinates	Ru
Urban	U
Water Area	W
Bare Rocks/Bad-lands	B
Others	O

N.B. (1) Smallest area; 5mm square representing 1.4 acre on 1:15,000 photos and 1.9 acre on 1:18,000 photos.
(2) Mixed crop pattern such as yams and banana: CA/Ba (CA 70% and Ba 30%).
(3) This list could be expanded if other major land uses are found.

PRESENT LAND USE MAP: HERMITAGE

5) Land use adjustment survey

Land use adjustment survey is essential for further planning of proper use and conservation needs.

Severely over-used lands should receive priorities for protection and/or adjustment. Under-used lands, especially those which are state or community owned, could be used for settlement or development purposes. Land which is currently used within capability limits may still need soil conservation treatments to ensure its perpetuity.

In obtaining such information, first consideration should be given to the land use policy of the government. For instance, whether or not natural forests (present use) on potentially cultivable lands are considered to be under-used lands and subject to intensive use or for settlement when required. Each use type should consequently be considered against each capability class to rate its degree of over-use, under-use and within-capability use.

After the proper criteria are determined, a map showing land use adjustment needs can be produced by overlaying a present land use map on a land capability map of the same scale (see Fig. 6). The list of criteria and a land ownership map should be kept on hand in the mapping process.

Example 7 shows the mapping procedures, a criteria list together with a sample map for reference.

The map shows the sites and areas of the use conditions. It only indicates the needs of adjustment; any adjustment and conservation plan should depend on further planning actions (see 8.3). Nevertheless, this map will make further development planning much more easy and objective.

Example 7

PROCEDURES FOR PRODUCING A LAND USE ADJUSTMENT MAP

1. Introduction

The land use adjustment map is produced by overlaying a present land use map on a land capability map. It contains information on use conditions such as over-use, under-use or use within capability but needing soil conservation treatment, etc. This kind of information is essential to the watershed planner for making decisions on future land use, adjustment needs, conservation needs, and/or possible resettlement (on public lands).

2. Criteria

A classification criterion for determining land use conditions was first worked out. This was produced by comparing present. land use with land capability to obtain the following results (see next page):

w+: Use within capability but needs soil conservation treatments.

w: Use within capability, soil conservation treatment not necessary.

o+: Seriously over-used. o: Over-used.

u+: Under-used public lands which can be adjusted for better use, i.e. resettlement.

u: Under-used.

3. Methodology

(1) Place the coloured present land use map on a light table.

(2) Place the transparent sheet of the land capability map on top the present use map and tape down firmly.

(3) Examine each parcel of land against its corresponding land capability to determine the land use conditions (w , w, o , o, u , u)

by using the above-mentioned criteria as a guide.

(4) For determining u+, a land ownership map should be superimposed atop of the two other maps.

(5) The smallest unit was about 1.4 acres (0.56 ha).

(6) Checking should be carefully done on the symbols and boundary drawings of each piece so classified.

(7) Make final mapping with ink and label properly.

CRITERIA FOR LAND USE ADJUSTMENT

Present Land Use	Land Capability: C1	C2C3	C4	P	FT AF	F	FP	Remarks
CA	W+	W+	W+	O+	O+	O+	O+	Annual crops
CP								Perennial crops
Cf	U	U	W+	O	W+	O	O	Coffee
Ba	W+	W+	W+	O	O	O	O	Banana
Ci	U	W+	W+	O	W+	O	O	Citrus
Su	U	W+	W+	O	O	O	O	Sugarcane
Co	U	W	W+	O	O	O	O	Coconuts
G	U	U	U	W	O	O	O	Fallow/Natural pasture, etc. If it is public land: U+
PI	U	U	U	W	O	O	O	Improved pasture
FF	U	U	W+	O	W+	O	O	Food forest
F								Forest
Pc	U	U	U	O	U	W	O	Plantation, Conifer
Ph	U	U	U	O	U	W	O	Plantation, Hardwoods
Fn	U	U	U	W	U	W	W	Natural Forest
Bo	U	U	U	W	U	U	W	Bamboo
Ru	U	U	U	U	U	U	W	Ruinate. If it is public land: U+

LAND USE ADJUSTMENT MAP: HERMITAGE

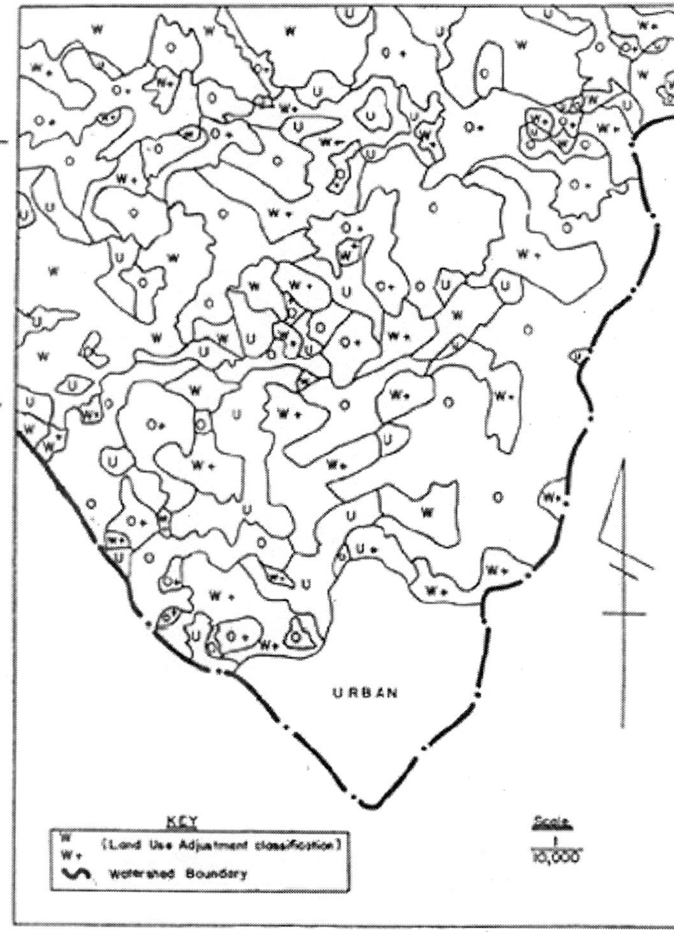

7.2 Erosion surveys

Soil erosion is a major watershed problem in many developing countries. In a watershed there may be many different sources of erosion. The main source areas should have been identified during preliminary investigation stages once identified, a detailed survey should be implemented using criteria and forms developed according to local needs. The main objectives are to pinpoint main erosion sites, define their extent, study their causes and, most importantly, suggest possible corrective or rehabilitation measures.

The survey should be management-oriented rather than academic-oriented. Air photo interpretation plus field checking should generally be sufficient for most erosion surveys. In the following sections five erosion-related surveys are briefly described:

- Geological survey and geomorphological analysis.

- Sheet and gully erosion survey.

- Road erosion survey.

- Landslide investigations.

- Survey of stream erosion and torrent.

1) Geological survey and geomorphological analysis

Geology and geomorphology of a watershed have significant indications on the fluvial processes of channels and hillslopes and erosion rates. In many countries, geological maps and information may be already available. However, the map scale may often be small and the information is not specific enough to cover the watershed in question. Some rechecking and refinement are usually needed.

If there is no existing information, a brief survey is required. However, the watershed manager should spell out what major data are needed from management point of view and not ask for a general survey which may include unwanted information and take too long a period to complete.

The basic geologic information needed is related to erosion and sedimentation. Rock types, depth of weathering, structures, among others, are the main concerns. For identification of erosion hazards and slope stability, further information should be provided on:

- severe displacement including faults, well-developed joints, extensive fractures, crushed zones, etc.;

- folded areas including anticlines, synclines, and homoclines and their strikes and dips, etc.;

- potential mass movement areas, including weak geological material, talus, recent deposits, dips parallel to slopes, and thin soils over impervious bed, etc.

Geomorphology deals with land forms in a watershed. The fluvial processes of channel development and hillslope evolution are the main concerns for a watershed manager. A survey of land forms will result in a better understanding of the erosion process, hazards, and hence of the management possibilities. For instance, a valley at youth stage will have more active erosion than one at old stage. Hummocky topography at the base of a hill is characteristic of landslide topography. High stream density usually means quick surface runoff and flash floods, etc. This kind of information, together with rock types and structures, permits proper selection of sites for dams and roads as well as estimation of peak flows and timing, etc.

In addition to collection of descriptive land form information, there are some quantitative analysis methods which can be used for comparison or interpretation. Fig. 8 shows some equations used for analysing morphological characteristics. With a reasonably good topographic map there should be no problem in implementing such an analysis for a watershed.

2) Sheet and gully erosion surveys

Sheet erosion data may be obtained in several different ways; the determining factors will be time, expenses, and existence of data:

- field investigations including interviewing farmers, checking fence sites and exposed tree roots and comparing soil profiles from adjacent undisturbed areas such as natural forest. Investigation should be carried out carefully and collect as much evidence as possible;

- collection of existing data including erosion data from soil survey reports and data from local experiment stations, etc.;

- setting simple plots with stakes and pins or with soil collection devices such as runoff plots for data supplementary or verification purpose;

- using soil loss prediction equations or erosion models. Care should be given to the validity of equations or models when they are used on steep slopes;

The survey of sheet erosion can be conducted in conjunction with soil survey. Erosion information can simultaneously be shown on the soil map. Example 8 shows a classification of sheet erosion by water used in the USA.

Gullies are comparatively easy to identify on air photos. However, field checking is still needed to ensure proper interpretation. Accurate measurements of gully development (head and channel) need bench mark setting and ground surveying. Whether this is desirable at the planning stage depends on time and resources.

Gullies can generally be classified by their stage (active or inactive), by their form (continuous or discontinuous), by their shape (V-shape, U-shape, etc.) and by their size (small, medium, or large).

Example 9 gives a classification of gullies by depth.

Figure 8

SOME QUANTITATIVE ANALYSIS OF WATERSHED MORPHOLOGY

Parameters	Equation	Remarks/Methods
Area (A)	$A = W \times L$	W: Watershed width L: Watershed length
Form Factor (F)	$F = \frac{A}{L^2}$	A: Area L: Watershed length
Compactness (C)	$C = \frac{2\sqrt{\pi A}}{P}$	P: Circumference A: Area
Stream Density (D)	$D = \frac{\Sigma l}{A}$	l: Total stream length A: Area
Stream Frequence (Sf)	$Sf = \frac{Sn}{A}$	Sn: Stream number A: Area units (e.g. km^2)
Mean Elevation (E)	$E = \frac{\Sigma e}{N}$	e: Total sample elevations (e.g. at map grids) N: Number of samples
Mean Slope (S)	$S = \frac{\Sigma s}{N}$	s: Total sample slopes (e.g. at map grids) N: Number of samples
Relief (R)	$R = H - Lo$	H: Highest elevation Lo: Lowest elevation
Relief Ratio (RR)	$RR = \frac{R}{L}$	R: Relief L: Watershed length
Stream Profile		Plot stream length against elevation
Hypsometric Curve		Plot elevation against area
Percentage Hypsometric Curve		Using percentages plot elevation (e/E) against area (a/A)

Example 8

CLASSIFICATION OF SHEET EROSION BY WATER

Class 1: Up to 25% of the original A horizon, or original ploughed layer in soils with thin A horizons, have been removed from most of the area.

Class 2: About 25 to 75% of the original A horizon or surface soil has been lost from most of the area.

Class 3: More than 75% of the original A horizon or surface soil, and part or all of the B horizon or other underlying layers, has been lost from most of the area.

Class 4: The land has been eroded until it has an intricate network of moderately deep or deep gullies. Soil profiles have been destroyed except in small areas between gullies. Source: U.S. Soil Conservation Service.

Example 9

CLASSIFICATION OF GULLIES BY DEPTHS

Description	Gully Depth
Small	3 feet (0.9 m) or less
Medium	3 to 15 feet (0.9 m to 4.5 m)
Large	15 feet (4.5 m) or more

Source: U.S. Soil Conservation Service.

Methodology of sheet and gully erosion surveys and studies can be seen from FAO Conservation Guide 1 (FAO, 1977), Hudson (1981) and Lal (1988).

3) Road erosion survey

Road erosion is a major watershed problem. Especially in mountainous countries, improperly constructed and poorly maintained roads often contribute large quantities of sediment to downstream areas through side-slope sliding and road foundation failures.

A survey of road erosion should concentrate on three parts: (1) cut slope, (2) fill slope, and (3) road surface and side ditches. During the survey, the site, problem and magnitude should be analysed and possible control or corrective methods be considered. At the end of the survey, the quantity of various control measures should be estimated. By multiplying each with a unit cost the total cost estimate can be worked out.

Fig. 9 shows the major causes and forms of road erosion and their control measures. Details can be seen in Sheng and Stennett (1975) and FAO conservation Guide 13/4. Example 10 shows a survey form for reference.

4) Landslide investigations

Landslides are mostly caused by natural factors, i.e. heavy rains, steep slopes, weak geologic formation, etc. However, in many countries, as development takes place in upstream watersheds, the hazards and damages of landslides also increase. Landslide prevention and rehabilitation therefore becomes an important watershed task.

A landslide is a downward movement of a land mass from a slope. The major landslide forms are fall, slide, slump, flow, creep and their combinations. Many classifications of landslides have been developed; some depend on material and movement, others depend on causes and still others depend on mechanics and age, etc. A local classification can be developed according to actual needs.

Whatever the classification system, it should be practical for watershed management purposes. The following are some points for consideration:

- the classification should be applicable to both natural and artificial slopes, and natural and man-made causes;

- it should facilitate application for both air photo interpretation and field investigation;

- the classification should provide basic information for treatment and rehabilitation needs.

Major survey activities should involve photo interpretation, field checking, recording and mapping. After considering treatment needs and corrective measures for each slide, a cost estimate should be produced.

Example 11 shows a landslide classification and investigation form based primarily on "immediate causes".

Figure 9

FORMS OF ROAD EROSION AND THEIR CONTROL MEASURES

Position	Erosion, Causes & Forms	Control Measures
I. Cut slopes	(a) Concentrated runoff	1.Diversion ditch forms gullies
		2. Waterway
	(b) Lack of protection	3. Sodding causes erosion
		4. Planting block, vegetation belt and hydro-seeding
	(c) Retrograded sliding due to sub-surface flow or seepage	5. Portable retaining wall.
		6. Sub-surface drainage
		7. Buttress
II. Fill slopes	(d) Poor drainage promotes gullies and foundation failure	8. Drainage improvement
		9. Earth dyke or berm
	(e) Tension cracks due to poor filling and stream cutting	10. Elimination of cracks
		11. Riprapping and toe filling
	(f) Surface erosion due to lack of protection	12. Wattling and staking
		13. Greening paper
III. Road surface and side ditch	(g) Steep road gradient causes track erosion	14. Cross drain
		15. Grass seeding
	(h) Steep side ditch induces bottom scouring	16. Ballasting, checks & vegetation

Example 10   ROAD EROSION SURVEY

Code	Road	Kind	Mileage*	Problem Area	Size	Cut	Fill	Sur-face	Drain
1	Treatment Plant/ Christiana Road	H/S	1.84	a) 1 mile from the plant b) close to bridge neer the plant c) 0.5 & 0.6 mi. from the bridge	5,280 ft 20 ft 15 ft 12 ft		 X X	X	
2	Christiana/Dump Road	H	2.16	a) 1.2 & 1.8 mi. from Chr' tiana b) Pumping Station	500 ft^2 600 ft	X			 X
3	Dump/Moravia Road	H	3.47	a) 0.7 to 1.4 mi. from the Dump b) 1.5 mi. from the Dump c) 2 mi. from the Dump	5,100 ft^2 1,200 ft^2 50 ft	X	 X X		
4	Zorn School (or Low Christiana Road)	S	1.26	a) 0.25 mi. from town b) 1.5 mi. from town	50 ft 800 ft^2				
4-1	End Trail	T	0.45	Up and down portion at streambank	600 ft			X	
4-2	Blue Hole Branch	S	0.34	Lower end	240 yd^2	X			
4-3	Tyne Town Branch	S	0.37	At the end	200 ft			X	
5	Sinclair Road	S/T	0.34	a) At the upper end	320 ft^2	X			

*Measured from 1:5,000 scale topographic map.

Example 11

EXAMPLE OF LANDSLIDE CLASSIFICATION AND INVESTIGATION F(

D
Ie

_________Watershed Landslide Classification and Investigation Form

Code No.		Sub-Watershed		No. of Map and Photo		Symbols_____
Location		Elevation	_____M	Aspect		
Site	Slope_____, Bedrock_____, Soil_____					Other information:
Volume	Width_____M, Length_____M, Depth_____M Size_____M², Volume_____M³					
Land Use	Dry farming(CD)_____, Terrace(T)_____, Rice paddy (CP)_____, Orchard (O)_____, Natural forest (F)_____, Tree plantation (FP)_____, Bamboo(B)_____, Road or highway (H)_____, Grassland(G)_____, Denuded, plantable (LP)_____, Denuded, unplantable(LU)_____, Brushland (J)_____,					Sketch or photograph:
Immediate Causes	Runoff Concen. (R)_____, Stream cutting(S)_____, Seepage (Sp)_____, Road excavation (Ex)_____, Cultivation (Cu)_____, Geologic causes(Ge)_____,					
Phenomena	Wash out (W)_____, Deposit (D)_____,					
Age	New landslide(N)_____, Enlargement(E)_____,					
Stability	Relatively stable(St)_____, Unstable(Un), _____ Highly unstable (V)_____,					
Treatment	Needed immediately(A)_____, Needed eventually(B)_____, Not needed(C)_____, Not justified or feasible(D)_____,					
Treatment Recommendations	Retaining wall etc.(Re)_____, Diversion (Di)_____, Bench terracing(Bs)_____, Wattling (Wa)_____, Reforestation(Rf)_____, Grass seeding(Gr)_____, Check dam(Ck)_____, Clearing(Cl)_____, Drainage(Dr)_____,					

Source: Sheng, 1966

5) Survey of stream erosion and torrent

Streams usually reflect watershed conditions and respond to major hydrologic events. When the natural equilibrium is lost in a watershed due to people's excessive activities or extremely heavy rains, the stream below will display significant bank cuttings, scouring or sediment deposition. While man can hardly control natural events, efforts to protect the watershed will normally result in less erosion in the streams.

Stream erosion surveys can also be done by air photo interpretations and field checking. In addition to watershed conditions, three items for immediate consideration are bank cutting, channel stabilization and sedimentation. Streams can be classified according to degree or seriousness of erosion in order to determine treatment priorities. Example 12 gives a classification system originally developed by the US Forest Service and modified to suit other countries where more serious stream erosion exists. For streambank protection and water quality control, sometimes a protection belt along streams is needed. To establish the belt needs a survey of present conditions and use. Example 13 gives the width required for the belt on various slopes.

In highly developed or heavily populated watersheds, torrential mountain streams often cause heavy damage to the nearby villages and downstream areas. These streams having steep gradients, extreme fluctuations of flows and massive bedloads are very dangerous and unstable if not controlled.

Techniques for torrent control surveys have been well developed and extensively used in Europe, Japan, and many other countries where villages, hotels, or recreation areas are situated in mountain watersheds. In wildland watersheds or at different socio-economic conditions of less developed countries this kind of work may seldom be practised due to the relatively high cost. Control measures should not only cover streams and adjacent areas where damages may occur but also include the respective tributaries where the torrents start. Control measures usually include vertical and horizontal channel stabilization, bank protection, control of tributary gullies and slides, and revegetation. Detailed torrent control survey techniques can be seen from a FAO publication "Torrent Control".

Example 12

EROSION CLASSIFICATION FOR STREAMS 1

Class 0: Stream shows no signs of erosion or excesive discharges. Banks well vegetated, often overgrown with woody vegetation. No recent debris or flotsam on banks or lodged in vegetation. Streambed formed of shingle or cobbles, weathered and often discoloured by algae. Pools well developed.

Class I: Signs of incipient erosion. Banks undercut and raw in places; fresh sand and sediment in pools; sand-bars active; streamside vegetation may be gone or disappearing.

Class II: Accelerated erosion evident, sand bars active, gravel and rocks scoured clean, pools filled with sediment. Streamside debris and flotsam deposited on soil and vegetation well above banks. Some tributaries gullied and depositing fans or deltas in main stream.

Class III: Severe erosion. Same symptoms as Class II. but aggravated. Stream bends cutting out actively; bottomland being lost by bank cutting; stream turbid or carrying bedloads most of the time.

Class IV: Very severe erosion. Extremely active bank slides and cutting; heavy deposits of fresh bedloads; torrent nature of flows.

---

1 A modification of the original classification by H.G. Wilm of the US Forest Service.

Example 13

RECOMMENDED WIDTHS FOR STREAM PROTECTION BELT[1]

Slope of Land		Recommend Widths[2]			
		Municipal Watershed		Ordinary Watershed	
Percent (%)	Degree (°)	Feet	Metre[3]	Feet	Metre[3]
0	0	50	15	25	8
10	5.7	90	27	45	14
20	11.3	130	39	65	20
30	16.7	170	51	85	26
40	21.8	210	64	105	32
50	26.6	250	76	125	38
60	31.0	290	88	145	44
70	36.0	330	100	165	50

1 Trimble, Jr., G.R. and R.S. Sartz. 1957.

2 The widths could be doubled under humid tropical conditions where rainfall intensities are very high.

3 Approximate conversions.

7.3 Water and other natural resources surveys

In any watershed, there exist many kinds of natural resources: water, soil, forest, range, wildlife, etc. Depending on management objectives, many basic surveys or investigations of these resources are needed from which better management plans can be prepared.

In the following sections, brief descriptions of five survey techniques relating to natural resources are given:

- climatic surveys;

- hydrological surveys;

- investigation of water resources and use;

- forest, forest land and agroforestry surveys;

- range, wildlife and recreation surveys.

1) Climatic surveys

Climatic surveys usually concentrate on items such as precipitation, temperature, evaporation, humidity and wind, etc. which either affect the water balance and erosion of the watershed or influence vegetation and crop growth of the area.

Most of the data can be collected from climatic stations in a watershed or from stations nearby. However, some compilation and analysis work is usually necessary. The basic information may include the following:

- precipitation: form, amount, distribution, intensity, etc.;

- temperatures: maximum, minimum, mean, frost days, etc.;

- others: evaporation, wind (speed and direction), humidity, radiation, etc.

If there is no such data available, new stations need to be set up. In addition to usual precipitation measurement, some stations may need basic instruments as shown below:

- a standard raingauge plus an automatic raingauge or recorder;

- a thermometer (for maximum and minimum temperature), a hygrothermograph (for continuous humidity and temperature) and/or a psychrometer;

- an evaporation pan and an anemometer.

Rainfall data is probably the most important factor relating to water resources, crop production, runoff and erosion. Three methods are usually used for estimating average rainfall in a watershed. They are 1) arithmetic mean, 2) Thiessen polygon (polygons are formed from the perpendicular bisectors of lines connecting nearby stations), and 3) isohyetal method. Example 14 shows how to apply these methods. For rainfall intensitites, data from automatic raingauges should be used and analysed. A hydrologist or a textbook should be consulted to obtain intensities at different time periods and frequencies. If the only data available are daily rainfall, the following equation can be used to find intensities of various short durations:

$$I = \frac{R}{24}\left(\frac{24}{t}\right)^{0.6}$$

where I = rainfall intensities, in mm/hr

R = maximum 24 hour rainfall of 10 year return period

t = duration or time of concentration, in hr.

The constant, 0.6, has been used for rainfall of 10 year return periods. This exponent can be modified to fit local conditions and for different return periods. A monograph and example can be seen from the same series of FAO Field Manual 13/3 (FAO, 1988).

Example 14

THREE METHODS FOR ESTIMATING AVERAGE RAINFALL FOR A WATERSHED

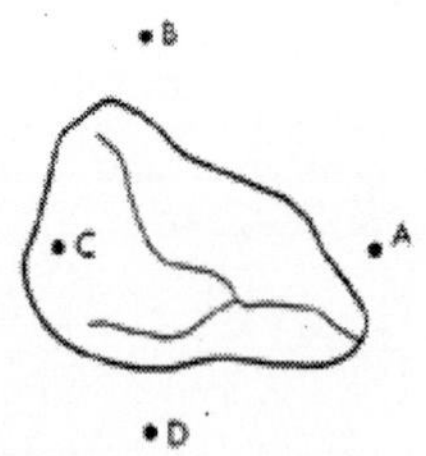

SOURCE DATA:

Daily rainfall measured at each gage in centimeters

A	B	C	D
4	8	10	6

Arithmetic Mean = $\frac{4 + 8 + 10 + 6}{4}$

= 7 centimeters

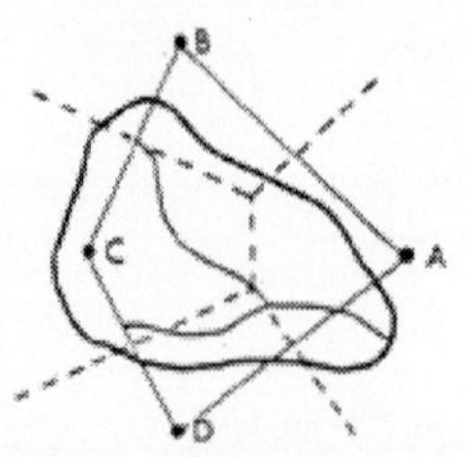

THIESSEN POLYGON:

Station	Depth (cm)		Area in Polygon*		Volume (cm)
A	4	x	.28	=	1.12
B	8	x	.09	=	0.72
C	10	x	.49	=	4.90
D	6	x	.14	=	0.84
			sum	=	7.6 cm

*As a fraction of total area

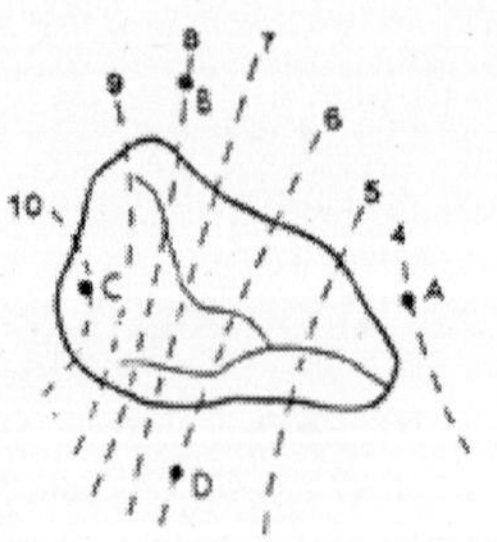

ISOHYETAL:

Mean Depth (cm)		Area Between Isohyets*		Volume (cm)
4.5	x	0.12	=	0.54
5.5	x	0.25	=	1.38
6.5	x	0.14	=	0.91
7.5	x	0.13	=	0.98
8.5	x	0.18	=	1.53
9.5	x	0,14	=	1.33
10.5	x	0.04	=	0.42
		sum	=	7.1 cm

Source: UNESCO MAB Regional Training Course

2) Hydrological surveys

For most developing countries, the important items in hydrological surveys are streamflow, runoff and sedimentation. Water quality may also be important in some countries.

Streamflow and runoff

The data required on streamflow and runoff are: a) peak or flood flow for designing engineering structures, b) low or minimum flow for estimating water supplies and c) annual total and its variation for various planning and design purposes.

Three major conditions may exist for such surveys. The first one is that there are already stream gauging stations established in the watershed with years of records. In this case, work is limited to compilation and analysis.

The second condition is that there are some gauging stations in downstream areas or in neighbouring watersheds. In this case, two methods can be used, given that the gauged watershed and the ungauged one under study are similar in climatic and biophysical conditions:

a) direct transfer: Using simple ratios of the areas of watershed to obtain needed data;

b) regional analysis: Using a statistical approach in which generalized equations, graphical relations or maps are developed for estimating the required information at ungauged sites. For more detailed information, "Water Supply Paper 1543-A" and "Regional Analyses of streamflow characteristics" of the US. Geological Survey should be consulted.

The third possibility is that no gauging stations exist in the same region and therefore no data are available. In this case the following analysis is recommended:

a) field investigation and estimation: For instance, visit the channel and look for and enquire about high water marks to estimate flood flows. This can be done by measuring channel profiles, calculating mean velocity by using the Manning Formula (see Field Manual 13/3), and estimating the corresponding peakflows using the following equation:

$Q = AV$ where:

Q = discharge in volume (e.g. cubic metres per second)

A = cross-section area, in square metres

V = velocity in metres/second;

b) measurement of flowing water, surface floats can be used for rough estimation of velocity. More accurate measurement can be done by using a current meter. On wide streams, many horizontal sections should be divided for individual readings. Once the average velocity is obtained, multiply it by the cross-section to get Q;

c) peakflow estimation: among many equations used, the Rational Formula is still popular in many countries where basic data is lacking. Especially for small engineering structures and local floods this equation should be considered appropriate:

$$Q = \frac{1}{360} CIA \qquad \text{where:}$$

Q = peakflow, in cubic metres per second

C = runoff coefficient, the I of rainfall appears as runoff (e.g. forest land: 0.2-0.3, etc.)

I = maximum intensity for a selected frequency in a duration equal to 'time of concentration", in mm per hour

A = area of watershed, in ha

For sample calculations, Field Manual 13/3 (FAO, 1988) can be consulted.

d) new gauging stations: Water level recorders and staff gauges need to be set up to gather data not only for planning purposes but also for future evaluation. Using natural control on the stream as gauging site, a stage-discharge curve (rating curve) should be eventually established for each station. If using artificial control (e.g. on small watershed or for watershed experiment) many formulas can be used. A standard textbook of hydraulics should be consulted;

e) watershed models: some of the popular models have been introduced in Appendix 3.

Sediment Survey

Sediment is usually divided into two classes: suspended load and bedload. Many methods can be employed for their surveys. Brief descriptions of selected methods follow:

a) Suspended load:

- suspended sediment in a stream can be measured by "sediment samplers". A popular sampler used in the USA for wading measurements is USDH-48. Six to a dozen samples equally spaced across the channel are usually sufficient for smaller streams. Current meter readings are normally made at the same time. For large rivers, measurements are usually done from a cable, a bridge, or a boat;

- for intermittent or ephemeral streams, a multiple stage sampler can be installed along the bank in order to obtain sediment concentrates at different heights of flow;

- for small watershed and plot studies, sample devices such as the "Coshocton Wheel" can be used. It is used in conjunction with an H-flume and takes a certain fraction of run-off for sediment analysis. Simple sediment tanks can also be used for plot studies.

Sediment in the samples can be filtered and dried. The dry weight is expressed as a concentration, in milligrams per litre or ppm (parts per million). To get total suspended load, sample sediment concentrations should be multiplied by stream discharge. A sediment rating curve for a stream should eventually be developed and used for estimation. For detailed methodology, a standard hydrology textbook should be consulted. Also, FAO Conservation Guide 1 and 2 (FAO 1977, 1976) provided some techniques.

b) Bedload:

Bedload is usually difficult to measure, especially on large streams. However, several ways can still be employed for measuring bedload in a small watershed.The simple ones are described as follows:

- for ephemeral streams, a trench across the channel can be dug to trap bedload. Cleaning and measuring need to be done after heavy rains or after filled up;

- behind a check dam, a siltation survey can be made periodically to compare the previous profile and the present one to estimate the rate and the amount of bedload. This type of survey can be applied to any structure where the impounding water is not deep;

- a soil loss or run-off plot will usually collect data on coarse sediment or bedload.

c) Total sediment deposits:

A survey of the cross-sections of reservoirs and ponds against original or previous profiles will provide figures on total sediment deposits over a given time span. A pole, a sounding cable or an automatic depth recorder can be used depending on water depths and budget. Normally, the reservoir management organization carries out periodic surveys and the information can be used for estimating the total sedimentation rate of a watershed. FAO Conservation Guide 2 has detailed descriptions.

3) Investigation of water resources and use

This investigation usually covers water budget, water use and use problems and possible solutions.

Water budget and resources

Each watershed is a hydro-morphological unit which responds to precipitation and energy inputs and produces streamflows and evapotranspiration as outputs. A water balance or water budget concept can be expressed simply in an equation as follows:

$Q = P - ET \pm S$ where:

Q = total streamflow or runoff including measured groundwater flow

P = total precipitation

ET = total evapotranspiration

$\pm S$ = change in storage

This equation represents a normal and balanced picture in which no serious leakage from or to the watershed is occurring. Starting with the driest or wettest month when the soil water storage is believed to be constant from year to year, a water budget of a watershed can be calculated. If all but one component can either be measured (P and Q) or estimated (S at the driest season approaching the wilting point), the application of the equation is relatively simple: $ET = P - Q$, assuming changes of S in a year period are very small. Potential ET can also be calculated using evaporation data from a pan. Example 15 shows water budgets for two places in Thailand.

Such findings, together with the hydrometeorological data, will give an overall picture of the potential and problems of the water resources in a watershed. For instance, is there sufficient water presently for all the uses? Or in what months is the shortage most severe and what can be done to alleviate it?

Water use

Water use includes domestic, irrigation and industrial use. Each one should be investigated as to conditions of present use and predicted future use. For forecasting future use, estimates need to be made on the growth of the population, cropping area, and industry, together with their respective use rates. An example of such surveys at two watersheds in Jamaica can be seen from JAM/67/505 Project Technical Report 13/1 (FAO, 1977a). A brief account of survey items is described under 7.4 (Infrastructure surveys).

Water use by vegetation through evapotranspiration can be estimated as mentioned before. Although applications are still limited, research results have shown that manipulating vegetation can increase streamflow.

Example 15

Average monthly water budgets for two stations in Thailand

Station: Chanthaburi, Thailand	OCT	NOV	DEC	JAN	FEB	MAR	APR	MAY	JUN	JUL	AUG	SEP	YR
							mm						
1. Average rainfall	231	74	15	48	46	76	117	325	498	444	439	478	2791
2. Initial soil moisture	279	279	238	136	75	18	0	9	215	279	279	279	--
3. Total available moisture	510	353	253	184	121	94	117	334	713	723	718	757	--
4. Potential ET	123	115	117	109	103	110	108	119	123	128	129	124	1408
5. Actual ET	123	115	117	109	103	94	108	119	123	128	129	124	1392
6. Remaining available moisture	387	238	136	75	18	0	9	215	590	595	589	633	--
7. Final soil moisture	279	233	136	75	18	0	9	215	279	279	279	279	--
8. Runoff	108	0	0	0	0	0	0	0	311	316	310	354	1399

Station: Chaing Mai, Thailand	OCT	NOV	DEC	JAN	FEB	MAR	APR	MAY	JUN	JUL	AUG	SEP	YR
1. Average rainfall	130	46	10	5	10	13	51	127	132	198	232	290	1244
2. Initial soil moisture	124	124	63	0	0	0	0	0	24	46	124	124	--
3. Total available moisture	254	170	73	5	10	13	51	127	156	244	356	414	--
4. Potential ET	114	107	102	99	85	87	87	103	110	117	120	112	1243
5. Actual ET	114	107	73	5	10	13	51	103	110	117	120	112	935
6. Remaining available moisture	140	63	0	0	0	0	0	24	46	127	236	302	--
7. Final soil moisture	124	63	0	0	0	0	0	24	46	124	124	124	--
8. Runoff	16	0	0	0	0	0	0	0	0	3	112	178	309

Source: UNESCO MAB Regional Training Course

Use Problems

Water use problems usually include water quantity (overall shortage, seasonal shortage, etc.), water quality and legal and economic aspects of water use.

The legal aspects of water use such as various use rights should be investigated. Related laws or regulations should be reviewed and, if necessary, suggestions should be made for modifying them or proposing new legislation. Institutional problems should also be studied and improvements suggested. The cost of using various types of water should also be reviewed because it affects not only use rates and the total water resources but also the well-being of the people in the watershed.

Water shortages of any kind should be investigated. The timing and degrees of seriousness should be examined and possible solutions suggested. These may include storing excess runoff during the rainy season in tanks, ponds, small dams; using additional water harvesting techniques; digging wells; practising moisture conservation measures on crop lands; or even manipulating vegetation to increase streamflow.

Water quality problems can be divided into those of chemical, bacteriological or physical origin. Sediment is probably still the most setious physical problem in upland watershed areas in the developing world. For chemical and bacteriological qualities, periodical grab samples should be collected and sent to water treatment plants or public health centres nearby for analysis. What watershed people should do is to identify the source or contributing areas. Possible solutions to such problems may include relocating feed lots, setting stream buffer stripes, and treatment of factory waste and mining tailings.

Example 16 shows the tolerable turbidity of water for various uses under USA conditions for reference purpose. A manual on monitoring stream water quality for land-use impacts (kunkle et a1. 1987) can be a good reference.

4) Forest, forest land and agro-forestry surveys

The management of forest and forest lands, in most countries, started much earlier than the management of watersheds. In fact, watershed management is a relatively new branch of many national forest departments. Most of the data on forest resources, management plans and forest protection needs should be already available.

Five surveys, however, will be dealt with under this section. They are: 1) forest roads and logging; 2) protection forest; 3) revegetation needs; 4) hydrologic conditions of soils, and 5) agroforestry.

*Example 16_*

TOLERABLE TURBIDITY OF WATER FOR VARIOUS USES

Water Use	Tolerable Turbidity (mg/l)
Drinking	5
Industry	
Canning	10
Cooling	50
Dark paper	25
Light paper	5
Textiles	5
Swimming	10
Boating	20

Source: Chang, 1982.

Survey on forest roads and logging

Forest roads and logging, if not properly operated, may cause serious problems in a watershed. Therefore, a survey of their conditions is usually needed. Example 17 provides sample check lists of them. This example should be used in conjunction with Fig. 9 and Example 10 in Section 7.2 (Erosion Surveys).

Survey of protection forest

Surveys or reviews of the needs of protection forest is often an important part of watershed surveys. In many countries, protection forest was established many decades ago when the socio-economic and technical criteria were very different from those of today. A careful review of the necessity of the existing forest should be undertaken. Usually, new criteria need to be established including consideration of impact on nearby communities and other social consequences.

Additional area or new protection forests are sometimes required in the headwaters of a watershed. Objectives should be well defined - for soil stabilization, water conservation, or flood control - because the required management practices are different in each case. Also, the functions and limits of protection forests should be made clearly known to the community nearby in order to avoid misunderstanding.

Whilst there are no universal criteria for determining the needs for protection forests, the major factors should include slopes; soil (depth and erodibility); geology (rock, structure, etc.); site (headwater, roadside, etc.); and socioeconomic and legal conditions. A point system may be developed to determine whether there is a need for this kind of forest in a watershed.

Survey for revegetation needs

A survey on revegetation needs should be carried out on any erosive or disturbed areas where protective vegetation cover is needed. The required survey should include:

- logged or cut-over areas;

- eroded areas in forest such as landslides, mined-out and deposit areas, eroded sites along roads and streams;

- special areas which need to be revegetated such as reserved areas, buffer stripes, abandoned squatting areas, over-grazed forest lands, etc.

Based on the survey, a revegetation plan should be prepared including methods (replanting, reseeding, or natural revegetating), species, nurseries, planting or seeding seasons, together with a work programme and cost estimates.

Example 17

SAMPLE CHECK LIST FOR FOREST ROAD AND LOGGING ASPECTS

I. Road aspects

1) Are roads built away from swales, valley bottoms, and elongated depressions?

2) Are roads built on benches, ridges, and toes of slopes where feasible?

3) Do roads avoid seeps, clay beds, slide areas, and steeply dipping formations?

4) Are there adequate buffer strips between roads and streams?

5) Are there natural stream channels altered only where the need is urgent, and only then in accordance with standards?

6) Do gravel and borrow pits contribute to soil displacement and stream pollution?

7) Are cut-and-fill material and debris permitted to reach stream channels?

8) Are road-clearing widths excessive?

9) Are cut-and-fill slopes of proper gradient for soil class encountered?

10) Are cut-and-fill slopes stabilized by mulching, planting, retaining walls, or other measures as necessary?

11) Are fill materials suitable to withstand wet season slump?

12) Are fills compacted by rolling during construction when necessary to ensure stability?

13) Are road surfaces contributing to stream silting by vehicles splashing muddy water into live streams?

14) Are road surfaces crowned to prevent concentration of water on road?

15) Is surfacing material spread over road surface rather than windrowed along edges?

16) Are fill slopes safeguarded from runoff by a protective berm (shoulder, ridge, or curb)?

17) Are culverts (or bridges) of adequate size and length installed at all stream crossings?

18) Are culverts installed on natural slope of the land with headwalls and footwalls as needed?

19) Are bog holes in roads drained promptly after discovery?

20) Is water diverted from roadside drainage ditches sufficiently often to prevent ditch scouring?

21) Are drainage ditches placed above cut slopes as needed to divert runoff from the slopes?

22) Are culverts and drainage ditches cleared of debris prior to rainy season and at other times as needed?

23) Are roads sprinkled and graded as needed during use?

24) Are trails contributing to silting of streams?

25) Are non-system roads closed to public travel and properly treated?

26) Is incompleted construction suitably drained at the end of the work season?

27) Is ditch-blading done so as not to undercut road slopes?

28) Do watershed management staff men review road plans prior to final approval?

29) Are specifications for erosion and water control in the road plan adhered to?

2. Logging aspects

1) Do logging plans specify areas of relatively unstable soil to be protected?

2) Are landings treated to prevent erosion?

3) Are sediment settling basins provided below landings where necessary?

4) Is use of tractors on steep and unstable soils causing erosion and stream siltation?

5) Are tractors used only when soil is firm enough to properly support them?

6) Are tractors kept out of stream channels?

7) Are trees felled or yarded across streams?

8) Do logging practices disperse rather than concentrate runoff?

9) Are skid trails located as far from live streams as feasible?

10) Are skid trails, landings, and spur roads properly drained?

11) Are skid trails and landings reseeded promptly where needed with soil-stabilizing species where this practice is advisable?

12) Are efforts made to scarify and loosen compacted soil in landings and skid trails compacted during logging?

13) Are temporary culverts and bridges removed at end of the logging operation?

US Forest Service Handbook 2533.

Hydrologic conditions of soils

Forests with undergrowth and thick layers of litter and humus are very different hydrologically from those without them. Also, different types of soil and different sites have different rates of infiltration. A special survey of the hydrologic conditions of forest soils is needed in many cases for watershed management purposes, especially when flood control or water supply is the main objective.

The survey usually concentrates on the following areas:

- density of the forest and undergrowth;

- kind, density, and thickness of the ground cover;

- soil textures and infiltration rates;

- site conditions including slope, elevation, erosion, etc.

Such a survey may be carried out to cover the soils of entire watershed. Example 18 shows hydrologic soil groupings based on infiltration capacity and texture. Example 19 shows the criteria used in the USA for determining hydrologic conditions of forest and woodland.

Example 18

HYDROLOGIC SOIL GROUPINGS

A. (Low runoff potential). Soils having high infiltration rates even when thoroughly wetted and consisting chiefly of deep, well to excessively drained sands or gravels. These soils have a high rate of water transmission.

B. Soils having moderate infiltration rates when thoroughly wetted and consisting chiefly of moderately deep to deep, moderately well to well drained soils with moderately fine to moderately coarse textures. These soils have a moderate rate of water transmission.

C. Soils having slow infiltration rates when thoroughly wetted and consisting chiefly of soils with a layer that impedes downward movement of water, or soils with moderately fine to fine texture. These soils have a slow rate of water transmission.

D. (High runoff potential). Soils having very slow infiltration rates when thoroughly wetted and consisting chiefly of clay soils with a high swelling potential, soils with a permanent high water table, soils with a claypan or clay layer at or near the surface, and shallow soils over nearly impervious material. These soils have a very slow rate of water transmission.

Source: US Soil Conservation Service.

Ex. 19 CHART FOR DETERMINING HYDROLOGIC CONDITION OF FOREST AND WOODLAND

1 inch = 2.54 cm

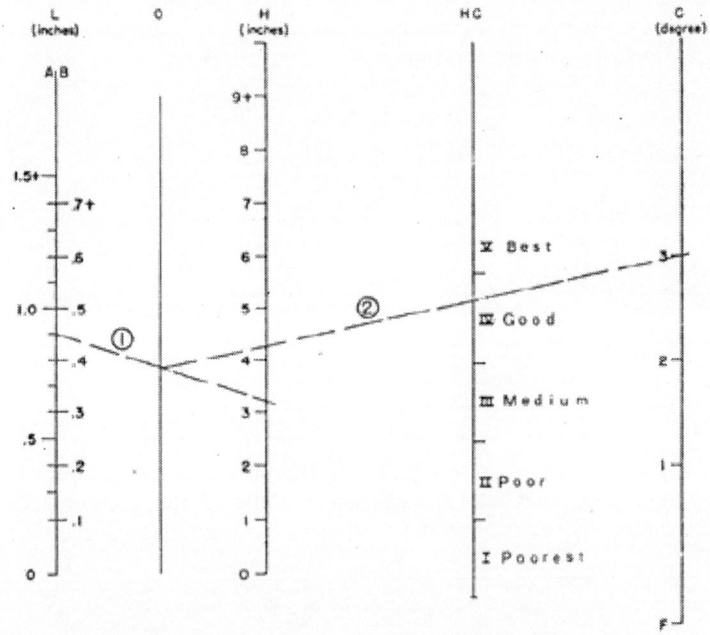

LEGEND

L– Litter depth
A– Litter before compaction
B– Litter after compaction
H– Humus depth
HC– Hydrologic condition class
C– Compactness
- F– Frozen, refers to "concrete" frost.
- 1– Compact, tight
- 2– Moderately compact
- 3– Loose, not compact, friable.

O– Turning axis

EXAMPLE

Given:

L = 0.9" (before compaction)
H = 3.2"
C = 3.0

Solution:

1. Draw line ① connecting L = 0.9 and H = 3.2.
2. Draw line ② connecting intersection of line ① and O scale, and C = 3.
3. On HC scale, read HC = IV

SOURCE: USSCS

Survey of agro-forestry needs

Although agroforestry practices have existed in many parts of the world for centuries, it is only recently that the subject has received special attention from forestry, agricultural, environmental and other scientific fields. Agroforestry systems, if properly designed and implemented, can benefit both small farmers and upland watersheds, especially where steep slopes have been misused over decades.

Agroforestry embraces many combinations of trees, agricultural crops and forage. There are also many systems of agroforestry, e.g. agro-silviculture, silvo-pastoral, agro-silvo-pastoral and multipurpose tree production systems, etc. The survey should first identify existing agroforestry activities in the watershed or in the country and examine their real benefits to the farmers as well as their impact on soil and water resources. In addition, survey of agroforestry needs should include the following:

- customs, capabilities and constraints of the farmers;

- appropriate system(s) for local needs;

- marketing possibilities and technological feasibility;

- possible environmental impacts.

Example 20 provides guidelines for surveying environmentally sound agroforestry projects.

Example 20

ELEMENTS OF SURVEY OF AGROFORESTRY PROJECTS

- Survey of needs, customs, and abilities of local people.

- Study of both existing and potential markets for future development.

- Examination of constraints related to economics, infrastructure, and the potential for organization of local community working groups.

- Decisions on which agroforestry systems would be most appropriate for local community needs, the ecological setting, and existing markets.

- Selection of management techniques, including planting and harvesting schedules, to maximize yields of both trees and farm crops.

- Provisions for monitoring of production and changes in soil fertility; this information should be used as feedback to improve the system.

- For intercropping, careful consideration must be given to the following:

1) optimum mixtures and spacing patterns of trees and farm crops, to maximize the production of both. (Particular attention should be given to complementary and conflicting relationships between species);

2) foliage characteristics and leaf fall of the various species, and their influence on competition for solar energy and nutrients;

3) shade tolerance of agricultural species and the effect of forest species on energy levels at the forest floor.

Source: Folliott, P.F. & J.L. Thames, 1983.

5) Rangeland, wildlife and recreation surveys

In addition to water, forestry and agriculture, many watersheds may have many resources that need to be surveyed and planned for better management or development. In this section, brief descriptions will be given to surveys of rangeland, wildlife and recreation resources.

Rangeland

The major task is to identify the main species of forage crops, measure their productivity and determine carrying capacities of the land. The general conditions of the rangelands in a watershed should also be surveyed in order to permit recommendations for improvements.

Clipped plots are often used for determining the weight of forage over a growing season. The utilization of the plants can either be estimated by percentage or by actual usage. Carrying capacity in terms of animal unit month (AMU) is then determined by the following calculation.

Carrying Capacity (AMU) =(forage produced) (proper use factor) / (forage requirement)

A survey of rangeland conditions also covers the degree of forage utilization, e.g. whether it is overgrazed or not; erosion damage or erosion risk; water supply; soil compactness, infiltration and environmental problems caused by domestic cattle, etc.

Wildlife

First, the wildlife population needs to be surveyed. This can be done by direct aerial counts, driving and counts, or the variable strip method. Indirect methods such as call counts for game birds, ratio methods based on marking of animals or kill data from legal hunting, etc., may also be used. The survey is not merely a count of total numbers but also a measurement of the composition of the animal population.

Second, productivity should be surveyed. Both the hunted and the non-hunted populations should be considered. The results will show the population trend including growth, mortality and diversity, etc. Animal habitats should also be studied, including analysis of extent and location of water and food, and covers for animals or birds. Clearly, any new developments in the watershed should avoid to jeopardize existing wildlife.

Recreation resources

Recreation resources are becoming increasingly important in many watersheds of the developing countries, especially those close to urban areas. According to a classic standard used by the US Department of Interior, six classes of recreation areas can be identified, based on their location, activity opportunities, level of development and administrative responsibilities, etc. These may be more applicable to developing countries than many recent classifications. Example 21 briefly defines these six classes.

Recreation sites should be analysed in terms of usage. Estimates should be made regarding the number of visits, length of stay, season of visits and activities, etc. The methodology for estimation includes self-registration, traffic counting and/or using samples and regression analysis.

An environmental impact survey or study may need to be included if intensive use of these areas is foreseen. This is even more important in municipal watersheds where water quantity and water quality are already in critical condition. The study should concentrate on possible damage or disturbance of soil, water, vegetation and other related resources at the sites and in downstream areas. The survey results will provide bases for prevention of camp fire, litter and waste disposal as well as pollution control needs.

Example 21

SIX CLASSES OF RECREATION AREA: A SUMMARY

High density recreation areas: Recreation areas near urban centers, having intensive activities, and requiring considerable investment.

General outdoor recreation areas: Areas easily reached from cities, offering extensive and vacation type of activities. Less intensive development than the first category.

Natural environment areas: Locations are rather remote, weekend and vacation type activities. Limited development.

Outstanding natural areas: Areas valued for sightseeing and study of natural features. Minimal requirements for development.

Primitive areas: Development not required or permitted.

Historic and cultural sites: Established locations, for sightseeing or studying. Development limited to preservation or restoration.

Source: US Department of Interior.

7.4 Infrastructure surveys

Infrastructure surveys are always needed for either integrated watershed development or for rural development type of watershed management projects.

Generally, such surveys include the following:

- transportation survey;

- housing survey;

- survey of domestic water supply, irrigation needs and energy;

- survey of public services;

- Survey of agro-industry.

1) Transportation survey

Existing road networks including highways, secondary roads, feeder roads, forest roads, etc., should be investigated. Airphotos may provide part of the needed information whereas other information, e.g. transportation services, traffic and new development plans, etc., can be collected from institutions such as local governments and transportation authorities. Further information should be gathered on road surface conditions, alignment, road gradient, stability of side slopes, drainage problems, and road use and maintenance. Attention should also be given to trails used by animals and farmers. Required improvements such as surfacing (asphalting, ballasting, marling), additional drains (culverts, cross drains), slope stabilization (retaining walls, seeding and revegetation, etc.), adjustment of alignment, gradient improvement together with maintenance practices, should all be included. Such a survey and the road erosion survey mentioned previously can be combined as one if necessary.

A study of present traffic conditions and predicted future trends may point out a need for new roads. If it is needed, a reconnaissance type of survey for the new roads is usually sufficient at this stage to show alternative routes, benefits, and approximate costs.

Transportation facilities and services provided by public and private sectors should also be investigated. The quality of services, transportation cost to the markets and farmer's preference, among others, should also be included.

2) Housing surveys

Housing surveys and planning provide basic information for either new housing schemes or for improvement of existing ones. Many countries may have existing reports on national housing conditions and statistics from population census or physical planning. Such data identifies current housing problems in rural areas as well as future needs according to population forecasts. Based on these data, a localized special survey can be designed. Or, questionnaires on housing can be included in socio-economic surveys.

The survey for rural housing usually includes types of structure; construction material (for frame, wall, floor, etc); age of the house; rooms per household; total area (sq. m); number of persons per household; ownership; utility; kitchen and kind of stoves; type of toilet; present status; maintenance; and the owner's views on improvement, etc.

Improvement of existing houses may be the major task in many watersheds. Plans fen putting up additional rooms and improving roof and kitchen and toilet facilities can be of great benefit to local residents. Simple designs should be included in the future plan.

3) Survey of domestic water supply, irrigation needs and energy

Water supply for domestic use is usually the primary concern of the inhabitants. The survey should be designed to identify the present water supply system, if any, and its capacity, delivery, distribution (in yards, into dwellings or at roadsides), and potential for development, etc. Use problems, e.g. quantity, quality, timing, charges, and possible improvements should also be covered in the study.

Where a supply system is insufficient, further investigation of springs, wells and other sources is required to provide a base for planning of new developments or improvements. Possibilities for use of concrete water basins and storage tanks, and applying water harvesting techniques, etc., should also be included.

Usually, minor irrigation and water harvesting can be planned by watershed people with some assistance from irrigation authorities.For irrigation needs, the survey should cover farming systems, cropping patterns, crop water requirement, source of water, distribution systems, energy needs, types of irrigation, and cost and benefit, etc.

An energy supply survey usually covers two main items, power and fuel. Public or private power companies should have statistics of present use and plans for future development. The survey needs to concentrate on the areas without electricity, their present light equipment and fuel and possible improvement.

Fuel for cooking and heating is always a major concern in upland watersheds. Women and children may spend most of their time collecting fuelwood and leaves, etc. On the other hand, the watershed manager is probably concerned about the destruction of forest and ground cover. Such a survey should, therefore, not only identify the kind and quantity of fuel consumed and its cost, but also search for alternative sources and means to alleviate the high labour input and vegetative destruction problems. The need for establishing village fuelwood plots at convenient places is, for instance, one of the subjects to be considered in the future plan.

4) Survey of public services

A general survey of public services in addition to water and electricity is always needed. Services directly related to agricultural production such as marketing, extension services, credit or loan facilities, etc., need to be investigated. Some of the investigations can be combined with socio-economic baseline surveys and institutional studies. Those local services which provide seeds, fertilizers, tools and pesticides, and storage facilities to the farmers through farmers' associations and others may need to be surveyed separately according to actual requirements. The basic information to be collected usually centres around the present service functions, their problems and potential, institutional adjustment needs and other improvement possibilities.

Services for general watershed dwellers may include schools, health clinics, post offices, community centres, etc. The quantity and quality of such services should be analysed and, if necessary, improvements should be recommended.

5) Survey of agro-industry

The possibility of developing agro-industry including cottage industry, merits, in many instances, a special survey. Especially when industrial and special crops are to be introduced in the watershed, a factory or a processing plant will surely induce farmers' production of raw material. Another important benefit of agro-industry is the provision of jobs or seasonal work to the watershed inhabitants including women and youth.

If a new agro-industry is contemplated, the survey should compile human resources data (labour availability, sex, age groups, unemployment, skill, education level, training needs, etc.) as well as physical resources data (water, power, land, raw material, etc.) required for the industry under consideration. In addition, an economic assessment and a financial analysis should be made in order to attract investors. Any potential investment sources should be identified and contacted.

Cottage industries are comparatively easy to start. Sometimes existing small shops can be improved or enlarged. Therefore, an investigation of existing shops and factories and their capabilities and potential is very necessary.

7.5 Socio-economic surveys

Socio-economic surveys are essential parts of any watershed survey and planning undertaking. Many technically sound projects have failed due to a neglect of socio-economic conditions of the watershed.

Five socio-economic surveys are discussed in the following sections. However, when funds and time pose a limit, one well designed and executed socio-economic base line survey should be considered sufficient in many developing countries.

- Socio-economic baseline survey.

- Demographic survey.

- Land ownership and settlement surveys.

- Survey of farming systems.

- Community development survey.

1) Socio-economic baseline survey

The subject of socio-economic survey confronts a vast array of social conditions and economic activities in a watershed. Usually, time does not allow for study of each sector in great detail. In many cases, a baseline survey is sufficient for collection of essential data for analysis. Periodical baseline surveys will help to identify the impact of project over time. Depending on project requirements, sometimes specific surveys or studies are further needed in addition to the baseline survey.

Before beginning a baseline survey, a series of decisions should be made on enumeration districts, listing, sampling methods, total samples, number of enumerators needed, survey methods, time and period of survey, data requirements, etc. In addition, important preparatory work such as the design of questionnaires, pre-survey and testing, and recruiting and training of enumerators, etc., should be properly carried out.

At economic part, the main objective of the survey is to collect data from farmers for farm development. Usually, three groups of data are needed:

- Identification of farms: head of the family; location; size; etc.

- Resources of farms: land (soils, slope, etc.); crops (food crops, tree crops, etc.); livestock (cattle, small animals, chickens); farm labour; farm machinery; implements and tools; irrigation water; farm buildings; financial liabilities (loans); and the flow of these resources.

- Utilization of resources: input/output of crops; input/output of livestock; material inputs to farms, general farm activities; off-farm activities; and household consumption, etc.

An FAO publication, "Farm Management Data Collection and Analysis" (Agricultural Service Bulletin No. 34, 1977b), describes such a survey in detail. Example 22 gives a farm summary table.

Example 22

FARM SUMMARY TABLE

FARM ANALYSIS PROJECT 786 HDR FARM FARM-MEAN-TABLE TOTAL AVERAGE SURVEY (54

COUNTRY	JAMAICA	DATA SOURCE	RESOURCE BASE
PROVINCE/REGION	UPPERYALLAHS	SOIL CONSERVATION AND	FARMER
VILLAGE/SUBREG.		WATERSHED DEVELOPMENT	FAMILY ADULTS 2.6
		SOCIOECONOMIC SURVEY	MAN-EQUIVALENTS 3.0
		FAO/ GOJ/	FARMLAND AREA OWNED 2.76
LONGITUDE	N18		CROPP PATTERN CROP(S)
LATITUDE	W077		
ALTITUDE.M	1000	AUTHOR: ECON. SECT. JAM/78/006	
SOIL TYPE	46/52 CL/SD LOAM YEAR OF OBSERVATION 1980		
CLIMATE TYPE	TROPICAL		TRACTORS 0.0

AGE 49.6 SCHOOL YEARS 7.3
PERM. LABOUR 0.1 YOUTH/OLD 0.7 CHI
RENTED OUT 0.0 RENTED 0.66 MANAC

CROP(S)	AREA	CROP(S)	AREA	CROPS	
28	0.15	29	0.11	37	
152	0.09	184	0.08	184	187
284 338	0.09	291	0.09	291	29
291 341	0.24	291	0.13	341	291
397	0.16	399	0.19	999	999

PACK ANIMALS 0.2 CATTLE 0.5 GOATS/SHEEP 1.1

CLIMATE	TOTAL	JAN	FEB	MAR	APR	MAY	JUN	JUL	AUG	SEP	OCT	NOV	DEC
RAINFALL MM	2319	132	117	76	114	203	163	107	173	224	409	376	226
TEMPERATURE MEAN	0.0	0.0	0.0	0.0	0.0	0.0	0.0	0.0	0.0	0.0	0.0	0.0	0.0
HIGH													
LOW													
HUMIDITY PERC													
LABOR INPUTS (MD)													
CROPS LAND. PREP.	30	3	2	5	5	3	3	2	3	1	1	1	1
PLANTING	9		2	1	2	1	1	1	1				
CULTIVATION	12	1	1	2	1	1	4	1	1				
HARVESTING	11			1	1	1	2	1	1	1	1		2
OTHERS													
TOTAL	62	4	5	9	9	6	10	5	6	2	2	1	3
LIVESTOCK ACTIVITY	48	4	4	4	4	4	4	4	4	4	4	4	4
GEN. FARM ACTIVITIES	48	4	4	4	4	4	4	4	4	4	4	4	4
TOTAL FARM	158	12	13	17	17	14	18	13	14	10	10	9	11
(OF WHICH FARMER	103	8	8	12	12	9	12	9	9	7	8	8	1
FAMILY	43	3	4	3	4	4	5	4	3	3	3	3	4
P. LABOUR	3		1				1		1				
S. LABOUR	8	1	1	1	1	1	1		1				1
OTHERS)	1			1									
OFF-FARM ACTIVITIES	180	15	15	15	15	15	15	15	15	15	15	15	15
POWER INPUT (HOURS)													
ANIMALS													
TRACTORS													
IRRIGATION WATER													

FARM ECONOMICS	CASH
FARM PRODUCTION	
CROPS	1051
LIVESTOCK	126
TOTAL	117
VARIABLE COSTS	27
GROSS MARGIN	90
VALUE SHARECR.	
GROSS MARGIN	90
OVERHEAD COSTS	
RENT	
TAXES	
PERM. LABOUR	6
INTEREST PAID	
REPAIR/MAINT.	8
OTHERS	
TOTAL	15
NET FARM EARNING	74
OFF/FARM EARNINGS	173
FAMILY EARNINGS	248
VALUE HOME CONSU	
HOUSEHOLD EXPEND	286
BALANCE	-38
PRODUCTIVITIES	
GROSS MARGIN/UNIT AREA (C	
LABOUR (MD)/UNIT AREA (CR	
GROSS MARGIN/MD CATTLE	
GROSS MARGIN/MD (CROPS)	
GROSS MARGIN/MD (CATTLE)	
GROSS MARGIN/ME (FARM)	

NET WORTH STATEMENT	FARM LAND	PERM. CROPS	BUIL-DINGS	TOOLS/ MACHINE	DRAFT ANIMAL	PROD. ANIMAL	TOTAL OWNED	TOTAL OWED	NET WORTH VALUE	PERC.
OPENING VALUES	3394	179	6913	85	122	345	11038	44	10995	99.6
INCOMING			4			86	91			
OUTGOING			11			65	77	5		
DEPRECIATION	3		313		12		328			
ENDING VALUES	3391	179	6593	85	110	366	10724	39	10685	99.6

For the social part of the survey, the data are needed for basic project designs and for determination of management strategies. If the main objectives of a project are erosion control and conservation farming, for instance, the data requirement may cover the following aspects:

- Basic information: age; educational level; family size; marital status; religion; etc.

- Establishment in farming: years in farming; attitude towards farming, land tenure; family members and time devoted to farming; farm income as percentage of total income; etc.

- Integration in the community: linking with the community; membership in community organizations; source of assistance; etc.

- Identification of farming problems: land too steep; soils too poor; not enough land, labour shortages; lack of credit; lack of technical know-how for improvement; poor marketing; or uncertain rainfall and weather; etc.

- Identification of major needs and priorities: domestic water supply; roads; health services; schools; better homes; marketing; storage; credit; better extension services; more land for farming; irrigation water; employment opportunities.

- Knowledge and attitudes towards soil erosion: understanding and observation of soil erosion; erosion damage on the farm; major causes of erosion; needs for erosion control; farmer's perspective on erosion control methodology; etc.

- Soil conservation skill and needs: knowledge about soil conservation; experience with and willingness to practice conservation farming; preference for certain conservation measures; kind of assistance needed (technical, financial or both); education and training requirements; possibilities for group action.

2) Demographic survey

Many countries undertake demographic surveys or a population census every ten years or so. If recent data are available, they can be used and analyzed without much additional effort. If the last census is out of date, supplementary surveys are needed, but the old census should be used as a base.

The main objective in collecting demographic data is to see the trends in population growth in the watershed. More people means more infrastructure needs such as domestic water, housing and schools, etc., and also means more land, more food, and more jobs. The survey should also identify the movements and composition of the population. These are important to labour supply, settlement needs, youth development, employment for women and other activities in the local community.

The essential data include, but not limited to, the following items:

- Growth trend: crude birth rate; crude death rate; rate of natural increase (per 1 000 persons); etc.

- Migration: out migration; in migration; net balance.

- Composition: sex; age grouping; education level; household size; etc.

- Employment: self-employed; employed; unemployed; employment by occupations, sex and age; number and rate of unemployment; etc. - Predication: population growth; number of persons per household; changes in age structure; unemployment rate and job requirement; etc. This survey can also be incorporated into the socio-economic baseline survey in order to save time and energy.

3) Land ownership and settlement surveys

Land tenure is an important socio-economic factor which greatly affects the farmer's decision regarding land use, land conservation, and farm development. In addition to the baseline survey mentioned above, a land ownership survey should be carried out to provide information for planning purposes. Ownership maps and data can be used for establishing priorities for land conservation and development. For instance, leased or settled public lands should be treated with conservation measures first to serve as demonstrations for private farms. Owner-cultivated farms may receive higher priority than rented lands, etc. Land ownership data will also provide information for land use adjustment and settlement opportunities (see 4.4).

Most of the ownership data should already be available in the government's land department, land valuation office or through local authorities. Depending on time and resources available, an ownership map for a watershed can be prepared with varying degrees of detail. If cadestral surveys have never been done, approximate parcel boundaries should suffice since the map is not intended for legal purposes. If there is no record or no time for sorting out individual parcels, dividing the watershed into two or three categories, e.g. public land (with various agencies), private land, and government settlement and leased land, should be adequate for watershed management purposes.A detailed boundary survey and tenure investigation is very costly and will not be required in most cases.

For a settlement survey, brief accounts of the history and the extent of old settlements are usually needed. Recent settlements in the watershed should be carefully examined. The reasons for settlements, infrastructure provided, size of lots, land use conditions, average incomes, refunds and payments, total cost and benefits, and settler's comments, among others, should be collected and studied. The results will be useful for new settlement planning in the future.

4) Survey of farming systems

Viewing farm as a whole system, farming systems survey should concentrate on constraints, capacities, farmers' attitudes and the interactions among different components of farming in order to suggest appropriate technology and acceptable improvements. Although it is closely related to ordinary farm management, the approach is not tied to maximization of returns, rather, to understand their systems and thereby suggesting improvements to fit farmers' needs.

For watershed management purposes, a stratified survey of farming systems may be adequate. The items may include the following:

- Systems: rain-fed; irrigation; grazing; mixed; subsistence; others.

- Site: highland; lowland; easily accessible; accessible; inaccessible; erosive; non-erosive; etc.

- Constraints: land; water; labour; age; capital; technical know-how; extension services; access roads; marketing; etc.

- Farming practices: cropping; livestock; irrigation; soil conservation and soil management; production input and output; etc.

- Interactions: between crops and animals; animals and trees; and crops and trees; etc.

- Farming hazards: yield variations; drought; flood; water and wind erosion; pest and disease; etc.

- Attitude, habit, and culture: advanced farmers; fairly advanced farmers; conservative farmers; food habits; work habits; spending and saving habits; degree of self-sufficiency; religious vs non-religious; independent; cooperative; etc.

- Farmer's perspective on improvement needs: more land; more water; more capital; better varieties; labour and capital saving practices; modern farming equipment; improved production efficiency; changing of crop patterns; etc.

5) Community development survey

As mentioned previously, the inclusion of some community development work in watershed projects can help to win local support.

From the socio-economic baseline surveys, some information can be gathered from the individual farmer on his or her major needs and priorities. However, in a local community, there are many non-farmers and community group members who are interested in the development of the community as a whole. An additional or supplementary survey is sometimes needed to cover the whole spectrum of the community. This can be done either by field visits or correspondence depending on actual needs. Example 23 shows data needs for community development.

Example 23

DATA NEEDS FOR COMMUNITY DEVELOPMENT

Name of village group: number of villages in the group;

- Number of households (farmers, non farmers, total);

- Population (No.) (male, female over 14 years, children under 14 years, total number);

- Living conditions (in numbers of households: good, fair, poor);

- Number of households by farm size (<0.5 ha, 0.5-1.0 ha, 1.0-2.0 ha, 2.0-3.0 ha, >3.0 ha);

- Ownership (farmowner, tenant, share-cropper, etc.);

- Fuel source in percent (collected, fuel forest, oil and others); - Livestock (No.) (draft cattle, water buffalo, dairy cattle, hogs, ducks, chicken and others);

- Water supply (No.) (sanitary piped, village wells, individual wells, river and others);

- quality of road connections (in village, village to fields, village to village, village to market, village to service centre, etc.);

- Schools and education

- location and distance (primary village school, district school, middle school, illiterate adults Z);

- Home and cottage industry (No.) (bag machine, rope machine, spinning and weaving tools, wood carving and others);

- Power tools (No.) (tiller, duster, sprayer, duster-sprayer, thresher, 4-wheel tractor, etc.);

- Market (name, distance); - Service centre (name, distance); - Main supply centre (name, distance);

- Public facilities (No.) (electricity, telephone, post office, medical centre, hospital, midwife, community centre, public yard, public garden, etc.);

- Home gardens - No. and size for (vegetables, fruit trees, fuelwood, etc.);

- Village cooperative (consumer, producer);

- Village warehouse capacity (square meters, cubic meters); - Village store (No.) (cooperative, individual);

- Home improvement (No.) (houses, kitchen, fence, farmyard, storeroom, farm buildings such as cattle shed, tobacco curing shed, silkworm shed, etc.);

- Organizations (farm improvement club, home improvement club, youth club, village association, credit union, etc.);

- Indebtedness (to government, money lenders, etc.);

- Other improvements (according to locality);

- Priority of improvements as given by village community.

Source: FAO, 1976b.

8. PLANNING APPROACHES AND BASICS

8.1 Planning in general

Survey and planning is a continuous process. Data collected from surveys are for planning purposes while planning can not proceed without sufficient supply of survey data. Therefore, to separate survey and planning is impractical. Especially in developing countries watershed surveys are not carried out for academic study. Rather, they are for management purposes. Fig. 4 at the end of Chapter 3 shows survey and planning as a series of continuous activities. A dotted line dividing "survey and data analysis" and "planning, monitoring and follow-up" is only for the convenience of discussion.

While survey approaches and techniques have been dealt with in the previous chapters, this chapter will discuse planning approaches and basics using various survey results. Economic assessment, alternative considerations, plan formulation, monitoring and evaluation will be explained in the following chapters.

8.2 Planning approaches

Planning, by definition, is to "devise detailed methods for doing, arranging and making something". For different things different approaches should be adopted. For instance, planning an engineering structure such as a bridge is very different from planning a watershed complex. Some useful approaches employed in watershed planning are explained as follows:

Bottom-up approach

Many watershed projects have failed because farmers and local communities were not involved in the planning process. Watersheds in developing countries are heavily populated by farmers. Therefore, any watershed plan will not be successfully carried out without their support or participation.

Several ways can be employed to involve farmers in the planning process. For instance, existing farmers' organizations can be included in the survey and planning body. Local watershed committees can be organized for planning and implementation purposes. Conducting individual farm planning or group farm planning with the farmers will obtain details on how farmers will use, develop and protect their farms. Involving villagers and communities for planning community forests, pasture, roads, and other infrastructural needs are also scopes of the bottom-up approach. During such planning processes government policy and farmers' needs can be fully discussed. For watershed plans to be useful and workable, they should be well understood and accepted at grass-root level.

Iterative approach

Planning is an iterative approach. Before a final plan is prepared, many studies, assessments, alternative considerations and revisions will have to be made. Generally, a preliminary or interim report should be made by gathering the results and reports of each team (see examples given in Fig. 3). After receiving comments from all the related sources, a review and revision period begins. The process may need to be repeated several times to find the best results.

Flexible approach

A final watershed plan is not like a blueprint of a bridge. A watershed plan should be considered as a starting point and should be kept under constant monitoring and adjustment. There are many reasons. First, project life may cover 10 years or so and many unpredictable things, caused by nature or man made, may happen during the period. New problems need new policies and techniques to cope with them. Second, watershed management is a complex task dealing with social, economical, cultural, legal, institutional, and physical problems of a watershed. Difficulties may arise during implementation and many times the original strategies and goals need to be revised. Therefore, learning by doing is a very important process hence any such plan should be kept flexible.

Flexibility means leaving rooms for future adjustment, modification, or revision. Consequently, a monitoring and evaluation process should be built into the plan for this purpose. This also means that the planned targets should be progressive, i.e. smaller at the very beginning and gradually expanding with the added experience.

8.3 Land use and conservation needs planning

For many watersheds in developing countries, planning for proper land use and soil conservation needs is a most important task. Land use planning is the process of evaluating land and use patterns, together with other physical, social and economic considerations for selecting or suggesting the best alternative uses. Land use planning can be carried out at many levels. For our purpose, two levels may be sufficient. One is to plan the watershed as a whole including all kinds of land and ownership. The other is planning at farm or community level. Since planning conservation needs should be based on present land use and land capability or suitability data, the results of these surveys should be taken into consideration as the basic data.

Planning at watershed level

At the watershed level, the major concern of watershed managers or planners is whether the land is properly used. Over-use of land usually causes soil erosion and land degradation while under-use may cause waste and social problems in many developing countries. Therefore, the sites, areas, degrees of misuse together with ownerships should first be found out. As illustrated in Fig. 6 and in Example 7, such information can be obtained by superimposing the present land use map and the land capability map. Whatever the capability criteria are, the main principle is that land should be used according to the capability. Any use beyond its capability should be prohibited or discouraged. Use below its capability, though allowable from a conservation viewpoint, is an economical loss subject to further adjustment.

Experience shows both over-use and under-use may exist in the same watersheds in developing countries. These phenomena may need to be corrected. After a land use adjustment map (see Example 7) and its statistics are produced, planning should be initiated by stratifying land ownership, degrees of improper use, and adjustment priorities. The general procedures are explained as follows:

- Seriously over-used lands should receive first priority of adjustment or protection. On private lands, further individual farm planning with the farmers should be carried out to study a mutual strategy to alleviate the problem. If new policy and/or incentives are needed, suggestions should be made in the plan. On public lands, this problem should be brought to the agency concerned for devising practical solutions. These may include resettlement or reallocation on under-used lands in the same watershed.

- Over-used lands should be adjusted or protected accordingly. After planned with individual land users, a time table should be worked out. - Under-used lands can be planned in two ways. If they are publicly owned, these lands can be used for settling those farmers who are now over-using their lands. Under-used private lands should be planned for better use according to farmer's interests and existing policies.

- Lands which are used within or according to the capability can usually be divided into two sub-categories. One category still requires soil conservation treatments and another does not. For instance, existing paddies, terraced lands and forest do not need any treatment. But many slopes classified as cultivable lands and under improper cultivation will still need major conservation treatments in order to minimize erosion hazards.

Another major consideration for the planners is what types of land use or crops are best suited for the watershed. This may be based on land suitability and involve studies of present crop patterns, farming systems, government policy, agro-climatic conditions, marketing, agro-industry, and investment opportunities, etc. Any existing or proposed plans for crop development, crop zoning, agro-industry, forest or pasture development by either public or private sector in the watershed should be well analyzed and considered in the plan. A land suitability map and related information (see 7.1) may provide a sound base for such work.

Planning at the farm and community levels

Government may introduce and induce proper land use and cropping systems, but it is the farmer who makes the final decision. Therefore, planning at the farm and community levels are necessary.

Farm planning is a joint venture between government and farmers to draw up plans on how their farms are properly used and protected. Government technicians, equipped with knowledge of land use and soil conservation and guided by policy, give advice to individual farmer while the final decisions of the plan should be made by farmer according to his or her interest, resources and ability. The plan thus obtained will be much more realistic than the top-down type of plan. A farm plan usually shows each parcel in the farm its physical feature (soils, slope, capability), use conditions (present and proposed), major conservation treatment (existing, proposed), time schedule for development and protection together with costs and benefits, etc. To draw such a plan for individual farms needs considerable time. The watershed managers or planners should be aware of the numbers of farmers in the watershed and the time needed for such planning. If they feel that time and resources do not allow for such planning at the project planning stage, alternatives should be sought. The alternative may include the following:

- Group planning, through farmers' groups and organizations.

- Classifying all the farms in a watershed into several models. Carefully plan for each model and using the results for estimating land use figures, cropping trends and soil conservation needs of the whole watershed.

-- Using the information obtained from the watershed level (explained in the previous section) make a preliminary plan leaving room for future refining at the implementation stage.

In all cases, information on individual farms is needed in the implementing stage. Planners should therefore make it clear when these individual farm plans or group plans Twill be required.

Likewise, plans for the use and conservation of community lands in a watershed are needed. These may include community pastures, village forest, and recreation areas. The community concerned and the land users should be involved in planning future use of these lands including improvement, development and conservation needs.

8.4 Planning for watershed protection and rehabilitation

In addition to conservation of cultivated lands, specific plans are usually required for protection and rehabilitation of various kinds of lands in a watershed. While actual needs are depending on watershed conditions, the following planning work may normally be required:

- *Forest protection and rehabilitation.* For forest protection, planning work may include provisions for fire lookout towers, firebreaks, fire suppression crews and equipment, warning systems, education meetings, and forest patrol needs, etc. The need for protection forest in upstream areas and agroforestry for cultivated slopes require careful planning. For watershed rehabilitation, a reforestation plan including goals, schedule, species, techniques, nurseries, and roads, etc. is usually required.

- *Pasture improvement and protection.* Such a plan is usually needed for public pastures and range lands including the work of reseeding, fencing, rotational grazing, control of the number of animals, supply of water and sheds, etc.

- *Gully control, stream protection and landslide rehabilitation.* These may include using both vegetative and structural means. Check dams, submerged dams, spur dikes, riprapping, diversions, channel clearing or reshaping, reseeding, establishing stream buffer strips are some of the rehabilitation work needed.

- *Road erosion control.* This is a very important work in watersheds of developing countries. The plan should aim at the protection of existing roads: their surfaces, ditches and cut and fill slopes including rehabilitation of land slips and road foundations. The required work may consist of hydroseeding, retaining walls, wattling and staking, cross drains and culverts and their proper maintenance.

- *Other protection and rehabilitation work.* Such as mining control, mined area rehabilitation, pollution control, stabilization of housing sites on slopes as required.

8.5 Planning for rural and integrated watershed development

In many watershed projects, several components of rural development are normally required. Depending again on the actual needs of a watershed, the development plans may include the following:

- *Irrigation and water harvesting.* Even in the humid tropics, there exists a pronounced dry season for several months. No crops can be grown during that period. Minor irrigation and water harvesting for the provision of supplementary water at the beginning of the dry season will help farmers to grow one more crop, thereby increasing their income. In some watersheds, domestic water has to be supplied or supplemented by water collection devices. The planning work generally consists of studying rainfall and crop patterns, analyzing water requirements, identifying source areas, proposing water delivery, storage and distribution systems.

- *Road development and improvement.* This kind of work is always needed although the road authorities should bear chief responsibility for it. Joint planning by road authorities and watershed people is necessary. Planning for new roads should include carefully consideration of their necessity, sites and future maintenance needs. Otherwise, improperly built and maintained roads may create more erosion hazards. Planning for road improvement should include all roads and trails in a watershed. The latter are usually neglected but they are used by farmers daily. Drainage, road regrading, surfacing, slope stabilization and better maintenance are some of the improvements to be carefully planned.

- *Housing and building construction.* Housing includes building new houses and improvement of existing ones. The former is usually associated with settlement or land allocation schemes and normally have set standards for planning and designing. However, watershed planners should help to select safe sites for housing and to review housing standards from both practical and economical points of view. Housing improvement may be needed extensively in many watersheds. Kitchens, toilets, and roofs may be the priority items for improvement. Other construction calling for planning are markets, schools, and clinics. As with housing, there may be competent authorities in charge of their respective development. Watershed planners and managers, however, need to assist with proper siting and site stabilization.

- *Other development plans.* Other development plans may be required such as establishing small power plant, cottage industry, agro-industry, etc. In each case competent authorities or experts should be consulted.

9. ECONOMIC AND OTHER ASSESSMENTS

9.1 The concept

Aspects of assessment

The major aspect of assessment work is economic assessment or appraisal. However, many other assessments should also be taken independently or in conjunction with the economic assessment. Consequently, assessments may cover seven aspects as follows:

- Technical aspects, including technical criteria, practicality, farmer's acceptability, technological effectiveness, etc.

- Managerial aspects, including institutional capability, coordination, extension services, training, etc.

- Economic aspects, including determination of investment justification for the project, identification of alternatives, and examination of project contribution to national and local economy, etc.

- Financial aspects, including examination of cash flows, returns to farmers and individuals, incentives, repayment schedules, etc.

- Commercial aspects, including marketing channels, supplies of material, arrangement of procurement, etc.

- Social aspects, including income distribution, employment opportunities, women and youth involvement, etc.

- Environment aspects, including soil stabilization, ecology, resources conservation, etc.

Not all the assessments apply equally to a project. Economic and financial assessments can be carried out together since they use similar input and output data. During economic assessment, detailed technical criteria and relationships together with managerial capabilities will have to be examined. Commercial assessment may little be needed if only subsistence type of agriculture is involved in a proposed project. For integrated and rural development type of watershed projects, social aspect of assessments should be much emphasized. Many of the impacts of a watershed project are also environmental and normally need to be analysed and assessed.

Objectives of economic assessment

The main objectives of economic and financial assessments (or appraisals) are to provide answers to the questions asked by decision-makers such as follows: (FAO. 1987):

- Are economic benefits greater than costs?

- What is the budget impact likely to be for the agencies and for private entitles involved?

- Will the project increase economic stability of the affected region? Will it have balance of payments impacts?

- Will the project be attractive to the various private entities (e.g. upstream inhabitants) who will have to put resources into the project to make it work?

Since economic and financial assessments are important and necessary to every watershed project, watershed managers and planners who are not economists by training should understand the assessment techniques and procedures, and thereby provide the necessary inputs to such activities. On the other hand, economists who assess watershed projects should also understand the unique characteristics of watershed work for better coordination hence better results.

9.2 Economic characteristics of watershed projects

Long term work and inter-generation benefits

Watershed work usually needs long-term and persistent efforts to obtain planned results. For instance, land use adjustments, soil conservation, or reforestation requires long-term investment, management, and maintenance. Yet the full benefits may only be realized after one generation or more.

These characteristics make project planning and economic assessment difficult. One is that benefits predicted may not be realized in the long run due to rapid changes of socio-economic conditions of a country. For instance, the benefits or establishing fuelwood forests may be greatly reduced when the rural people change stoves. Another is the use of discount techniques to value future benefits. For example, using a 9% discount rate, , a $ 1 000 benefit 20 years from now is only worth $ 178.4 today, and for 30 years only $ 75.4.

The long term nature of watershed projects always puts these benefits unfavourably when compared with other projects in economic terms, unless economists and planners get together to identify fully their benefits to justify such projects.

Multiple use and multiproducts

Watershed projects usually relate to multiple use of watershed resources e.g. soils, water, forest, grass, fishery, mineral, etc. Use of one kind of resources may affect others. Increasing benefits of one type may impair or damage others. For instance, timber harvesting may increase erosion and sediments and impair fishery and recreation values of a watershed. On the other hand, the products of a watershed project can be many: food crops, fruits, fodder and livestock, fuelwood, timber, animals, and fresh water fish. Each of them is a specialized field.

For effective assessment, not only the production function, demands, price structures of each product need to be understood, but also the relationships and conflicts of resources use should be fully comprehended. Taking into consideration one aspect and forgetting the others, or counting only benefits and neglecting its negative impacts are not uncommon and can be misleading the decision-makers.

Externalities

A watershed project generates a host of effects which are not accounted for in the analysis of financial or economic impacts because they occur outside the market and do not directly affect the project cash flow. These are commonly referred to as "indirect", "non-market", "spillover" effects or "externalities". These effects should be included in the analysis of watershed management projects, because they affect the whole economic and environmental system, although not the cash flow of the project in question.

In many cases externalities are difficult to identify and quantify. Many of them cannot be meaningfully valued. However, when important, an attempt should be made to describe them at least in qualitative terms, if quantification and valuation are not possible.

An important point to remember is that when a positive externality has been identified, it is also necessary to search carefully for any corresponding negative externalities.

There is not one way to proceed in the identification, quantifaction and valuation of externalities. A great deal of experience is needed as is a good knowledge of the complex interactions between watershed management activities and their effects outside the project. Some externalities can be accounted for in the economic analysis of projects through shadow pricing inputs and outputs.

Spatial distribution of costs and benefits

Somewhat related to but not equal to externalities is the spatial distribution of costs and benefits of a watershed project. For instance, the effects of soil erosion alone may cover three areas (Sfeir-Younis, 1983):

- Intra-farm effects, e.g. loss in fertility, decrease in area cropped, decrease in cultivation intensity, reduction of soil depths and crop production.

- Inter-farm effects in a watershed, e.g. silting low lands or wetlands, sedimentation or eroding of upstreams, increase or decrease of runoff, decrease of groundwater supplies.

- Inter-area or downstream effects, e.g. sedimentation of river systems outside the watershed, siltation of reservoirs, clogging of irrigation canals, and increasing the probability of flooding.

Difficulties in quantification and valuation

Because of lack of basic data in many developing countries, quantification of benefits may constitute primary difficulties to any economic assessment. It is not enough, for instance, to say that the project will reduce flood damage. A series of data need to be collected such as flood frequency and magnitude, areas affected, heights and time of inundation in relation to crop and property damages, and the flood reduction effects of the proposed project. Without these, valuation of benefits is not possible.

As mentioned before, not all the benefits derived from a watershed project can be satisfactorily quantified and valued. Reducing losses of human life, improvement of living conditions, or increasing aesthetic beauty, etc. can hardly be put in terms of dollars and cents. Many attempts have been made (Hufschmidt _et al., 1983) to put values to these benefits. However a universal application of their methods remains to be developed. In many cases, a qualitative description of these benefits may satisfy the decision-makers.

9.3 Functions and limits of economic assessment

Watershed managers and planners need to understand the functions and the limits of economic (and financial) assessment and what an economist can do to help in project planning.

The main functions of economic (and financial) assessment are as follows:

- To analyse the project worth in the context of national economy using cost and benefit analysis techniques.

- To determine whether the project is feasible and attractive to investors e.g. farmers, government agencies, banks and others.

- To examine risk factors of a proposed project using sensitivity tests and to see whether the project will stand for changes, physically or economically.

- To assist project design and to present various alternatives to the decision-makers on scale, technology, timing of a proposed project.

On the other hand, watershed managers and planners should realize that although economic assessment is a useful tool to project planning it does have its limits. The major ones are as follows:

- As mentioned earlier, not all the watershed benefits can be quantified and valued. This makes cost and benefit analysis difficult. Sometimes qualitative descriptions are allowed.

- Economic assessment is a tool. It is as good as the data put in. Lacking basic data in many developing countries such as erosion and sedimentation rates, erosion and productivity relationships, and flood damage records may affect greatly the assessment results.

- Finally, economic assessment is only one of the many needed assessments of a watershed project though it is an important one.

9.4 Major techniques in economic assessment

Watershed managers and planners who are not economists by training need to comprehend the major techniques used in economic assessment for the sake of better communication, understanding and coordination.

The following sections are brief introductions of the assessment techniques using plain language. Detail knowledge can be obtained from many publications (FAO, 1980, 1987; Gittinger, 1972; Shaner, 1979; and UNIDO, 1972).

With and without project approach

Comparing the impact of "with" and "without" project is a normal technique to determine a project's worth. Their difference is the net additional benefit arising from the proposed project. However, this is not the difference between "before" and "after" the project because the "without" case is not static. For instance, erosion and land degradation will be going on over time without soil conservation measures ("without" project). Fig. 10 shows the with and without project conditions.

Discount

Project costs and benefits can only be compared at the same point of time. One hundred dollars 20 years from now does not represent the same value as one hundred dollars today. Therefore, the main objective of discounting is to bring the future values, benefits or costs, to the present values for comparison using an appropriate discounting rate.

Cost and benefit analysis for project worth

Cost and benefit analysis is an important technique to systematically compare the streams of costs and benefits in order to determine economic efficiency of a project or project worth. There are basically three measures using the same input data and assumption, as follows:

1) *Net present value or net present worth (NPV or NPW)*

This measure is used to determine the present value of net benefits of a project i.e. the difference between the present value of the stream of benefits and the present value of all the costs. A project (or certain component of a project) may only be accepted if this difference is zero or positive (B - C 2 0). To compare several alternatives, the analysis results can be ranked for decision making.

2) *Benefit and cost ratio (B/C ratio)*

This measure is to determine a ratio using present value of all the benefits in the numerator and the present value of the costs in the denominator. A project is considered to be economically sound or acceptable when the calculated value is larger than or at least equal to 1 (B/C > 1). The results can also be shown as a ranking of alternatives.

Fig. 10

RELATIONSHIP BETWEEN WITH AND WITHOUT PROJECT

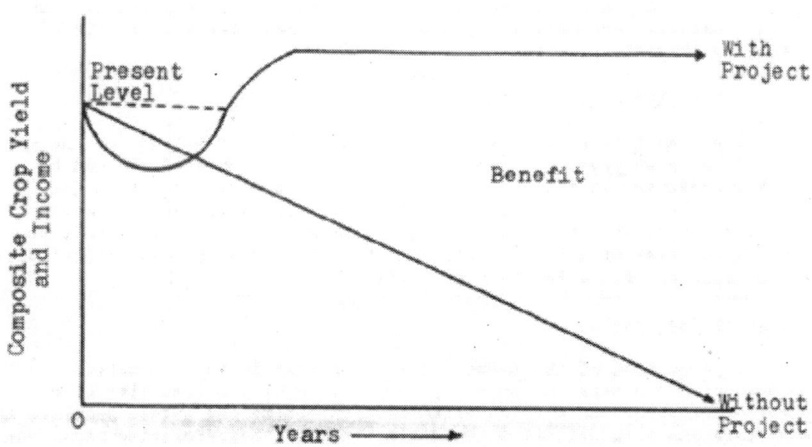

(Modified from Barlowe: Land Resources Economics)

3) *Internal rate of return (IRR)*

This is the discount rate, which, when applied to the stream of benefits and costs, produces an equal present value of both or a net present value of zero (A discount rate when B = C, or B - C = 0). This particular rate is called IRR and represents the average earning power of the project's investment to be compared with other investments.

Each measure has its pros and cons. NPW shows the magnitude of the net benefit of a project but indicates nothing about returns per unit. B/C ratio and IRR, on the other hand, give no indication of the magnitude of net benefit. Since they use the same set of data and a microcomputer can now help the analysis, economists may use all three measures with little effort to obtain a complete picture.

Cost effectiveness analysis (CEA)

When benefits have problems to be quantified or valued, or the work must be done by whatever the reason, cost effectiveness analysis can be used. The objective is to find out least cost to achieve the given goal (FAO, 1987). Or, costs are estimated in association with various levels of physical benefits and the results are presented to the decision-maker to decide which level of cost is justified (Shaner, 1979). Fig. 11 illustrates the principle of cost effectiveness analysis.

Sensitivity Analysis

It is a test of the impacts due to changes in cost, benefit, discount rate or others on the net present value or profitability of a project. Because of the many uncertainties in estimating future benefits and costs, this analysis will find those elements and their relative magnitudes of change that will affect net benefit of a project. Once identified, planners can then change the design of the project, build contingencies, or adjust the decision criteria (FAO, 1987).

Financial analysis

A project benefiting the nation as a whole may not be beneficial or feasible to the individuals who will invest in it, whether the individual is a farmer, businessman, private or public agencies. An apparent example is erosion control work which may sometimes benefit society more than the farmers in the upstream watershed who need to invest heavily in soil conservation work. As mentioned previously, financial analysis uses almost the same basic data as the economic analysis to study additionally cash flows, financial returns to the private entities, etc. in order to know whether the project is attractive.

Fig. 11

COST EFFECTIVENESS DIAGRAM

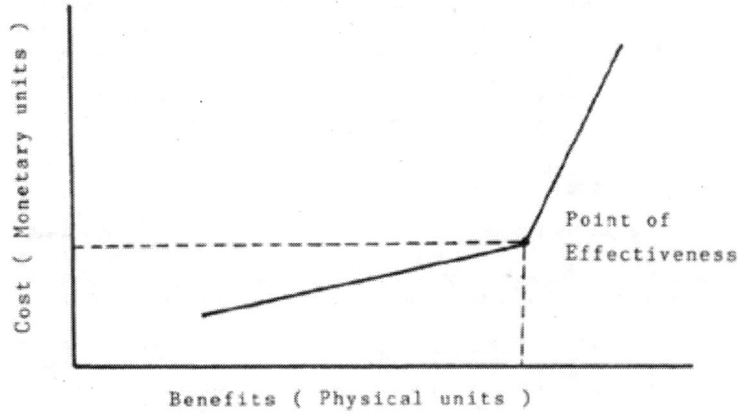

(Modified from Shaner, 1979)

9.5 Economic assessment procedures

An introduction

Planners or watershed managers should first realize that economic assessment or appraisal usually goes through a progressive and iterative procedure. Economists should be involved in the early stage of the survey and planning process for the purpose of initiating design, establishing technical relationships, recommending data needs, and analysing preliminary results. As the planning work proceeds, rough assessment and estimation of costs and benefits will have to be carried out many times, back and forth, to consider among alternatives. After the preliminary results put into the project's interim report, more work still needs to be done for revision or refinement as government authorities or funding agencies state their requirements.

Major Procedures

The major procedures involved in the economic (and financial) assessment are briefly explained here. Details can be seen from the Guidelines for Economic Appraisal of Watershed Management Projects (FAO, 1987) and the Public Management of Forestry Projects (OECD, 1986).

1) Developing technical relationships and quantifying physical inputs and outputs. The work includes defining and quantifying "without" project situation such as erosion and sediment rates over time; estimating "with" project impacts; and developing tables which show inputs and outputs for one or more alternatives.

2) Determining values and developing value flow tables. The procedure includes assigning monetary values; valuing watershed management inputs such as labour, equipment, material; valuing watershed

project benefits such as crop increases, dredging costs avoided, and flood losses reduced; and developing an overall value flow tables.

3) Measuring project worth. Using the techniques described in Section 9.4 to calculate and compare cost and benefit of each alternative.

4) Sensitivity Test. It is essential to test the sensitivity of the chosen measure(s) of project worth in terms of alternative values for many key factors, e.g. the discount rate, benefit value estimates, cost assumptions, etc. The results can be used for project revision or refining.

9.6 Assistance required for economic assessment

In the process of economic assessment, watershed managers and their team technicians need to work closely with economist(s). Not many economists have sufficient knowledge about the techniques of watershed management and soil conservation, and most of them will need technical inputs during the assessment of a project.

Assistance in establishing technical relationships

This is probably the first important step toward a sound assessment. The watershed technicians should try their best to provide technical information such as erosion figures on various land use types, sedimentation rates, effectiveness of various conservation measures in erosion reduction and crop production, and so forth. Such data should best be based on local experiment results and surveys. In developing countries where research and basic data are lacking, professional judgement, or information from similar ecological regions may be used. Realistic estimation should also be given concerning the watershed situation if no project were to be carried out over time. Likewise estimations of the expected results of the project should be realistic. Exaggeration or insufficiency will mislead the assessment.

Assistance in identifying costs

Costs of various watershed rehabilitation or protection practices are relatively easy to identify. Labour, equipment, raw material and land (purchasing or renting) can be accurately determined by experienced technicians and provided to the economist. However, the economist, after investigating the market price, may for certain reasons use different set of values such as opportunity cost and shadow price to reflect the real situation. The technicians should make every efforts to understand the underlying reasons for such assumptions.

Indirect cost may not be easy to identify. Assessment of watershed project impacts should also look into indirect cost and negative impacts to show the real picture. For instance, in some countries, gravel reduction in downstream areas may cause high construction cost in the cities due to material stopped by check dams in the upstream watershed. Such cost should be included in the assessment calculations.

Assistance in the identification of benefits

Insufficient identification of benefits has been one of the major reasons why watershed projects receive low priorities among other development projects. On the other hand, it is often difficult to identify, quantify, and value watershed project benefits when indirect and intangible benefits are involved.

Although much of the work will be done by the economist, watershed managers and technicians should understand some of the techniques used in quantification and valuation. Once understood, they can assist economist better to identify benefits. The following techniques are used quite often in the benefit valuation:

1) *Increase value of production:* Including crops, animals, timber, fruits, fish, etc. for greater physical production, quality improvement, better timing or better prices, etc.

2) *Avoidance of losses:* Such as dredging costs avoided, water treatment cost avoided, flood losses avoided, transportation losses avoided, and water use losses avoided, etc. A simple example of the latter is shown in Example 24.

3) *Cost savings:* Such as lower transportation cost due to new roads or better roads, less maintenance cost of hydro-power plants, and saving replacement cost of dams and equipment, etc.

Example 24

AN EXAMPLE OF ESTIMATING WATER SUPPLY BENEFITS

- Cave River/Pattoo Gully Municipal Watershed -

This benefit could be divided into two parts, benefit from non-stop pumping of water, and benefit from increase of streamflow. The first one can be estimated by cutting down about 80 percent of the pumping stoppage since very heavy rains may still make the streamflow unsuitable for treatment. The annual benefit, therefore, can be calculated as follows:

Ba = R F P

Ba = annual benefit in dollars

R = Reducing stoppage, i.e. 80 percent in a year, in hours

F = hourly pumping rate, in 1 000 gallons P = consumer price for 1 000 gallons

If we treat 1973 as a normal year, the total stoppage time is 302 hours. Eighty percent is 241.6 hours. Presently, the pumping rate is 30 x 1 000 gallons per hour. If the price of 1 000 gal. is J$ 0.60, then the annual benefit derived from this item is J$ 4 348.80. This is based on the present capability of the Plant. The benefit will increase with the increase in capacity in the future. On the other hand, the benefit for alleviating the discomforts of being without water cannot be estimated in monetary terms.

The other benefit of the watershed work is to increase the low flows during dry seasons. Vegetation cover and various kinds of terraces will provide excellent opportunities for infiltration of water and recharge of groundwater. It is also believed that with more water going into the ground rather than running off during the wet season, the low flows in the dry season could be increased. Although no figure could be given to the amount of increase without research findings in this country, any increase of low flows, or even sustaining the present level of low flow during critical periods, will benefit society a great deal.

J$ 1.00 = US$ 1.10

Source: FAO JAM/505 Project Working Document (1974).

4) *Increase value of land and properties:* Due to both improvement of environment and increase of productivity.

Three important things should be kept in mind in identifying and valuing watershed project benefits. First is to avoid double counting. Second, the negative impacts of a project may easily be neglected by enthusiastic technicians. A fair economic assessment should be based on all impacts. Last but not least important is that not all the benefits can be put into dollars and cents. As mentioned previously, such benefits should also be clearly described in qualitative terms and included in the final assessment.

9.7 Other assessments

In the process of economic assessment, many other aspects will have to be included or implemented such as technical aspects, managerial aspects, financial and commercial aspects, or social aspects, etc. Depending upon project needs and manpower availability, these assessments can also be done independently.

Experience shows that since many watersheds in developing countries are inhabited by small farmers, the social aspects of a project usually merits an in-depth and separate assessment. Emphasis could be placed on many facets of social concerns such as who will be really benefited from the project? Will the small farmers and rural poor be better off? What are the employment opportunities for women and youth? And how many upstream inhabitants take part in the decision making on watershed resources use?

Environmental impact is also a growing concern. Fortunately, watershed management and soil conservation projects are usually good for the environment. However, when rural infrastructures are included in a project such as road building, settlement and housing, water diversion, etc., their environmental impacts need to be assessed. Sometimes, environmental assessments are also required on extensive reforestation proposal in a municipal watershed, large scale forest conversion and crop development, and drastic land use changes in a watershed.

10. CONSTRAINTS, ALTERNATIVES AND STRATEGIES

10.1 Identification of constraints

The main purpose of identifying constraints is to consider alternative courses of work, counter measures, and proper strategies for better management.

Budgetary constraints

In many developing countries, budgetary constraints are often very serious. The common phenomenon is an absolute insufficiency of funds. Another kind of constraint is the uncertainty of forthcoming funds even when the budget has been approved. It is not unusual that field workers start work, sign agreements, promise subsidies to farmers in order to complete a seasonal task, yet they have to wait embarrassingly to get the needed funds. It is also not uncommon that funds allocated for watershed work are diverted for unrelated fields.

These constraints can be both discouraging and damaging. Planners, therefore, should identify them with other problems during the planning stage and make every effort to ensure necessary budgeting and proper flow and use of funds.

Manpower constraints

This is probably the most important constraint of all. A lack of funds can be solved by getting loans or grants from aid agencies or banks but work has to be done by people, especially people in the field.

Without properly trained technical staff, any project will fail. To identify the needs of technical staff, the following factors must be considered:

- Numbers, levels, basic qualifications and sources of staff.

- Training needs including type of training and time schedule.

- Incentive needs for field staff especially for those stationed in remote areas.

Watershed work is often labour intensive e.g. afforestation, terracing, gully control, etc. Although there may be idle hands in the uplands, they are often difficult to find at the needed time. Temporary migration to do a seasonal job such as cutting sugarcane, picking coffee beans or working at factories may cause labour shortages in upland watershed areas. Identifying such constraints will help to design implementation schedules and the establishment of realistic goals for the project.

Constraints relating to mobility and equipment

In developing countries it is not unusual to see that many of the best trained technicians are kept in the office and do little field work due to lack of vehicles, per diem, or necessary equipment. This is a genuine waste of time and precious human resources. Watershed work is a field-oriented task. There is no substitute for field work regardless of how well the plan is prepared.

The lack of vehicles in many countries has greatly hampered work progress in the field. It is very difficult to transport instruments such as surveying levels and rods by bus or by other means of public transport. Besides, many upland watershed may have no public transport. Unless the proposed project can provide sufficient vehicles, ensure that most of them are stationed in the field, and can make provisions for proper use and maintenance, this will be always a serious constraint.

In addition to vehicles, the proposed project should provide for sufficient per diem and travel expenses. If equipment, vehicles and instruments are to be imported from abroad, advance planning is necessary often, a project is delayed simply because vehicles and equipment have not yet arrived.

Constraints relating to technical information

Watershed management is a relatively new field in most developing countries. Few universities or colleges offer formal training in this subject, and research is also in its early stages. Therefore, technical information is always lacking.

If there are books or publications available, they are often written in foreign languages and contain information related to a different set of environmental conditions. For instance, watershed experiment results have been largely obtained from temperate zones whereas most developing countries are situated in the tropics and subtropics. Transfer of technology is a serious constraint and should be considered early in the planning stage.

Constraints in farmers' participation

As emphasized before, farmers' participation is a key to success in most of the watershed projects in developing countries where uplands are heavily populated with small farmers.

From a government point of view, there may be no reason to believe that farmers will not participate in a project if there are sufficient resources to help them. However, this may be over optimistic. The farmers themselves usually face many constraints which hamper them from participating in such a project, for example:

- They may not sufficiently comprehend the objectives of watershed management. They may feel that the government is asking them to protect the watershed for the benefit of others (downstream people).

- Traditional practices, for example, shifting or slash and burn cultivation, may not easily be changed over a short period.

- A conservative attitude may tend to resist any innovative or drastic measures.

- Shortages of labour and capital may restrict them to participate in any improvement task.

- Their economic status may not allow them to take any risks.

- They may be more interested in getting quick returns from their lands than in conserving soils for future use.

Whatever the constraints, the planners should identify them clearly with the help of socio-economic surveys, rapid rural appraisal, etc. Solutions or strategies should be sought and necessary arrangements be made to alleviate these problems.

Policy constraints and others

Serious policy constraints on land use and management should be brought to the attention of the government. For instance, lack of policy on encouragement or incentives for proper land use will result in difficulties in the land use adjustment previously described. Conflicting policy on use and management of various resources in a watershed may make implementation of watershed plans difficult. This with other institutional constraints should be well identified.

10.2 Management alternatives

Management alternatives should be studied and prepared during the planning stage in order to:

- Provide alternative courses of action.

- Keep plans flexible to cope with unforeseeable changes.

- Give government an unbiased look at the problems and their solutions

The best time to consider alternatives is when the field survey data are being gathered and analysed and the preliminary or interim report is being written.

Alternative work, costs and effects

Whether there are any better alternatives than the proposed work in terms of cost and effect is always a challenge to planners. Such alternatives may include a different approach, different kinds of work, changes of time schedule, location, etc.

If time permits, these alternatives should be evaluated systematically. The objective is to see whether or not a proposed change will yield increased benefits which are greater than increased costs, or the same benefits with reduced costs. For instance, for land protection and erosion control on steep slopes, allowing voluntary vegetation to grow may be the alternative to reforestation in the humid tropics. The latter will usually cost more and, if unsuccessful, will cause more erosion.

Technological alternatives and research

Research which identifies ways in which better results or lower costs can be achieved may support the adoption of technical alternatives. For instance, in soil conservation work, hillside ditches, a series of narrow and discontinuous benches, have been effective in erosion control (reducing erosion about 80 percent), requiring only one-fifth of the investment compared to bench terraces (which reduce erosion 90 to 95 percent). Unless farmers contemplate irrigation and mechanization which necessitate bench terraces, hillside ditches can be a valid technical alternative in protecting cultivated slopes.

Alternative budgetary sources

As discussed earlier, budgetary constraints are one of the major obstacles of adequate planning and successful implementation of watershed projects in developing countries. Most governments are interested more in rapid increases in production, earning foreign exchanges, and developing industry and cities than in protection of upland watersheds. Some alternative budgetary solutions that link watershed work with other development efforts are suggested below:

- A small percentage of funds from major construction or development projects in the watershed such as reservoirs, mountain highways, forest roads, mining, and housing development, can be allocated for protection purpose, especially for those activities which cause instability of watershed slopes.

- A portion of the earning from export crops such as coffee, banana, tea, pineapple, citrus, etc., which are grown on the slopes of upland watersheds and which need soil conservation treatment can be earmarked for protection purposes.

- Small fees can be added to the utility bills of dwellers in cities or towns which benefit from upstream watershed protection. The money can be directly distributed as incentives to the upland farmers who adopt prescribed conservation measures or used for watershed protection activities.

- Watershed or conservation districts may be established in some developing countries and fees and grants can be collected for watershed protection and improvement.

All the above-mentioned alternative sources need government policy or legislative support. However, because they are linked to other important development efforts, governments may be more likely to grant support in this manner than budgets exclusively for watershed protection.

Incentives and education

Whether or not farmers should be given material incentives (in addition to technical assistance) to adopt watershed conservation measures is a question which should receive much attention.

On the one hand, people realize that upland farmers are usually poor and should not bear the full cost of erosion control which will accrue benefits to others, e.g. reducing sedimentation and flood damages downstream. On the other hand, poor developing countries often cannot afford large expenses in the form of subsidies. Furthermore, farmers may develop a dependence on government handouts. Proper education may be more effective in the long run.

There is no easy answer to this dilemma. In general, some incentives are needed until farmers are convinced of the real benefits of such work (e.g. increased production and income). The length of time before these incentives can be reduced or eliminated will depend on extension efforts and farmers' income conditions.

There are many alternatives regarding the types of incentive and how they can be offered:

- Cash subsidies are relatively easy to handle and can be distributed during and after the completion of the prescribed work. Further decisions, however, should still be made on appropriate rates for each type of work; too much or too little will affect the outcome. Another important decision must be made on whether subsidies should be given for maintenance work and if so, for how long?

- Sometimes, to avoid misuse of cash, food, fertilizers and tools can be given. Payment in kind instead of cash, however, creates problems of purchasing, storage, transportation, and farmers' preference. It, therefore, adds burdens to the administration as well as to field officers unless such service is already available (i.e. under FAO World Food Programme).

- The government may directly hire crews to do some of the work of a more technical nature as incentives, e.g. waterways and gully structures on private farms. The potential problem is that when the local farmer does not participate in the work from the beginning he or she may assume that the government will be permanently responsible and therefore pay no attention to their maintenance.

- Exemption of tax on property and income can also be used as incentives. Adoption of this method requires an in-depth study in collaboration with the land and tax authorities to determine the proper criteria and period.

- Supervised credit or low interest loans with a reasonable grace period can also be used as incentive or as additional help to farmers. The effectiveness of these however, depends on the willingness of the farmers to incur debts, their capabilities to repay and whether they are qualified for the credit or loans.

- Other direct or indirect incentives including marketing and transportation arrangements, price support, and improvement of community amenities can also be considered.

Even if incentives are given, extension or education efforts cannot be allowed to idle. Short-term incentives may be used to boost farmers' enthusiasm and their participation in the beginning, but the success of any watershed project depends on the farmers' real understanding and their continuing support.

Project vs. programme

One important consideration facing any planner is whether the watershed needs a project to be carried out in a definitive time frame or needs a continuous permanent programme. If, for the time being, a project is preferable, what kind of follow-up plan is needed?

In many developing countries, watershed projects are supported by international funds over a period of several years. All too often, however, efforts are discontinued at the end of the project term and a new project with different funding is started somewhere else. This kind of "artificial injection" without continuity may cause negative effect. Discontinuity will not only damage government credibility and farmers' trust but also cause non-maintenance of roads, plantations, structures, etc., which, once failed, may induce more damage than before. Therefore, it is the managers' or planners' responsibility to explain and convince the authorities the need for a long-term approach to watershed management. If a foreign-aid project is proposed for the initial stage, they should also plan government follow-up activities and include the required long-term commitment in the proposed plan.

10.3 Strategies

Each country has its own problems and own conditions. Therefore, the following strategies are only for general reference.

Strategies for strengthening field implementation

Field implementation in developing countries is often hampered by many constraints. Strategies for strengthening field implementation which should be considered during the planning stages include:

- Establishment of sufficient field offices at strategic locations for accommodating staff working in the field. Necessary amenities should be provided so that the staff will not suffer when they are dispatched to the field.

- Provision of incentives such as special allowances (or hardship allowance), fellowships and better career opportunities should be made to field staff in order to encourage on-going work.

- Special achievement allowances are offered by many countries. The allowance is either given annually according to areas treated in conservation extension or extra per diem is paid according to the progress of work done in the field.

- Vehicles and equipment should be provided as needed. Field work should receive priority in allocation of vehicles and any abusive use should be prohibited.

- Coordination among various agencies should be ensured by better liaison and division of labour. Any conflicts or duplication of duties should be addressed and corrected in the shortest possible time.

- Budget funds and supporting services from head offices should be streamlined to back up the field operations.

- A field inspection, evaluation, and reporting system should be established at headquarters or regional offices to supervise and control the progress of work.

Strategies for transfer of technology

Transfer of technology is usually needed in watershed projects including information and experience from foreign countries, from other regions of the same country, as well as from technicians to the farmers.

For information collection, transfer and monitoring, a proper unit or post should be established to perform the following duties:

- Directly collect information and data from selected foreign institutions and translate or outline the ones having immediate interest.

- Liaise between national institutions on exchanges of information and data on watershed management.

- Systematically establish a data base for the use of technicians and farmers.

Foreign experts can be hired if needed and if there are resources to employ them. Fellowship abroad should also be considered. After returning, their final reports should be distributed and discussed among staff concerned. Seminars, workshops and training courses should be scheduled as part of the proposed watershed project. Through these activities, international experts, local specialists and returned fellows can share their knowledge and experience with others.

In order to transfer information and experience to farmers, result or process demonstrations on both public and private lands should be emphasized. Unless farmers can read, extension pamphlets may not have much use in upland watersheds. Practical training of farmers' leaders or contact farmers including visiting tours is very helpful because farmers usually trust neighbours more than outsiders.

Strategies for financial control

It is not unusual that a project or programme stops short of completion because of over-spending or because the original budget is insufficient to cover increased costs.

Although outside factors such as inflation, devaluation, or increases in minimum wages are difficult to control, the planners should prepare strategies to deal with these situations, should they occur. Some strategies are:

- Include an inflation factor in the cost estimates.

- Set up contingency funds in the project for unforeseeable future expenses.

- Exercise strict control of expenditures.

- Order or purchase equipment, vehicles, or material promptly as soon as the funds are approved.

- Endeavour to reduce costs through improved work efficiency and other means.

- Try alternative technology through research or field experiments to achieve the same or better results at less expenses.

Strategies for ensuring farmers' participation

Farmers' participation in protecting and developing watershed lands is a key to success.

Many strategies should be considered in order to ensure farmers' participation on a continuing basis. The following strategies are provided as examples:

- At the beginning of the project, small demonstration plots should be established in sufficient numbers on private as well as public lands to show the real benefits of the planned improvements, e.g. conservation farming.

- An intensive education and extension campaign should follow, using the results of the demonstration plots and experience of the farmers who participated in the demonstrations.

- If needed, a financial incentive programme (subsidies and/or credits) should be ready to help those farmers who are ready to participate in the proposed scheme.

- A technical assistance programme should also be available to whoever wants to join the watershed project. This is particularly important. Once interest is generated among the farmers there must be a programme to help them to plan and start the work; otherwise their enthusiasm will soon fade.

- A special effort should be made to organize interested farmers into neighbourhood self-help groups. The leader of each group can be designated as the contact farmer. He will receive intensive training and, if possible, partial wages from the project and will act as a bridge between the government and the local communities. He will also share his training with the others, and thus supplement the usually insufficient agents and over-worked extension service.

- A regular follow-up and inspection system should be established to help the farmers in maintenance, cropping and marketing activities, etc.

Strategies for proper maintenance

Maintenance is an extremely important part of watershed work but it is often neglected, with a resultant decrease in efficiency and increase in damage and waste. Strategies for proper maintenance should be carefully considered when the project is formulated.

On public lands and for public work such as reforestation, roads and check dams, budget provisions should be made for routine maintenance by government hired labourers. For watershed conservation work on private lands, a small incentive or a portion of the subsidies should be given for maintenance until the structures are stabilized or until plantations are established. An annual competition with small awards is another way to encourage farmers to continue proper maintenance. A sound inspection system should be established to oversee the maintenance activities.

Experience in some countries has shown that farmers, for the sake of cash subsidies, undertook ambitious soil conservation or tree planting work which they could not maintain. To avoid this kind of mishap, cautions should be taken as suggested below:

- For planners, a realistic target for protecting and treating farmers' lands according to their capabilities will be more fruitful than an ambitious one.

- For field officers, proper maintenance inspection should be considered as one of their major responsibilities.

- For farmers, treating or planting more lands than they can maintain should be realized as a waste of energy and time.

As mentioned earlier, for cultivable land after conservation treatment, follow-up or parallel services such as cropping, credit and/or marketing arrangements are extremely important. If the land remains unused or idle, for whatever the reason, the conservation structures or any improvement measures will usually not be maintained.

11. PLAN FORMULATION, RECOMMENDATIONS, AND MONITORING AND EVALUATION

11.1 Plan or project formulation

Determining management goals

After various surveys, analyses and alternative plans and considerations, realistic goals for the proposed plan or project should be determined. These goals are usually set against possible resources, institutional capabilities, government intentions and local needs. The goals should be progressive and allow for future adjustments.

Determining priorities and sequences

Not all watershed work can be started at the same time. There must be a logical sequence for implementation. As mentioned previously, priority sub-watersheds should be selected according to their locations, degree of disturbance, accessibility, management readiness and feasibility.

Priorities must be set not only for sub-watersheds but also for work. For instance, should the community road be built at year one or at year three? Should the farm ponds planned for the area be constructed before or after the road is built? And if after, how will the erosion and sediment from the new road affect those ponds?

For the integrated development of a watershed, the setting of priorities is a complex task. Should the project concentrate first on improving farm productivity or on development of infrastructure? Should a processing plant be established first or should it be built only after sufficient raw material has been produced? What are the priority needs of the local people? And what are priorities of the government?

A clear determination of priorities and sequences of work at the planning stage will benefit not only future implementation but also budget allocation. An orderly supply of funds is only possible with good and careful planning.

Preliminary or interim reporting

At approximately the mid-point of project planning, a concise preliminary or interim report should be prepared and presented to the government or other authorities for policy guidance, comments, and feedback. The interim report often contains a general assessment, alternatives and intended targets against possible investment and the time horizon.

This kind of preliminary report may omit many technical details but should emphasize important issues in addition to presenting the basic findings. The main aspects to be highlighted are:

- Explain long-term and short-term targets and their respective investment requirements for government consideration. If, initially, an international agency will support a short-term project, the government should be asked to continue the programme afterwards or make a long-term commitment. In the case of a purely government programme, the report will alert the government to the magnitude and time of investment required and enable it to make necessary preparations or commitment.

- Interpret existing government policy and, if necessary, propose new policy or its revisions to facilitate future implementation.

- Show the result of economic assessment.

- Obtain general approval of the planning methodology from the government.

Detailed financing plans

After the government and funding agencies have agreed on the new project in principle, financing is the final important facet the planners must confront. Without sufficient and proper financing, an economically sound project may never be started.

There are at least three parts of financing that should be well considered and planned. The first part is the direct investment from the government including funds for administration, training, fellowships, purchasing, materials, labour, transportation, etc. Although it is straightforward, it still needs proper and detailed budgeting.

The second part is the money used for production, usually in the form of loans to either private or public enterprise (e.g. agro-industry). Planning of this part is much more complicated. Cash flows, interest rates, returns and repayment schedules, etc. should be well planned.

For watershed projects, detailed financial plans are also needed for small farmers. For instance, the kind and size of credit, mortgage requirements, grace periods, repayment abilities and schedules, and subsidy needs, should all be studied thoroughly.

Determining the contents of final report

The contents of the final report should be determined as early as possible and not at the last minute. For a large and complex project, the best time to set the contents for the final report is after the interim report has been presented and preliminary approvals from all the authorities are obtained. For a smaller or simpler project, the contents can be determined earlier, immediately after the collection and analysis of data.

The contents of a final report will vary according to management objectives and actual needs. It is difficult to suggest a universal format. The following is a broad list of contents for general reference:

- Summary and recommendations.

- Descriptions of watershed conditions (biophysical).

- Analysis of major watershed problems (biophysical, socio-economic, institutional, etc.).

- Watershed management needs (including goals, alternatives, strategies, and effects, etc.).

- Economic and other assessments (including benefit, cost analysis and others).

- Work programmes (including targets, work schedules, budgeting, financial arrangements and monitoring and evaluation needs).

- Detailed recommendations.

- Appendices (including methodology, techniques, maps, photos, detailed figures, etc.).

Preparing the final report

The preparation of a final report is the last step of the whole task. To avoid delay in its production, once the contents are determined it is necessary to make concrete and detailed decisions concerning:

- The approximate length of each chapter or section.

- The person or organization responsible for preparing each chapter. - A deadline for draft submission and discussion.

- Nomination of a chief editor and specification of his or her duties. - A final deadline for report submission.

The scale of the various maps and the size of drawings and pictures also need to be decided upon in order to avoid unnecessary delay. The chief editor should keep in close contact with all the persons who are involved in preparation of the report. Any doubts regarding the format or contents of the report should be resolved without delay.

11.2 Recommendations on implementation

In addition to budget and financial matters, recommendations on how the proposed project can be effectively implemented are important parts of the plan. The following items are often included in the recommendations.

Responsibilities and operation mechanisms

The responsibilities of each organization which will be involved in implementation of the watershed project should be clearly defined. In addition, a field operation mechanism should be established to streamline the implementation. Following are some suggestions for reference:

- Liaison meetings should be held periodically at the field level and should be attended by representatives of all participating agencies. Ideally, the chief agency for planning or the liaison officer will act as chairman of the meeting. Thus a close link between planning and implementation will be established.

- Each organization should delegate responsibility to its field office or representative in order to permit smooth operations and work efficiency. Only important policy matters should be brought up to respective headquarters for decision-making.

- A joint supporting unit of administration and accounting can be set up in the field or attached to some field office nearby in order to speed up procedures of local purchase, disbursement, and field arrangements.

- In some cases, it may be appropriate to establish an autonomous body or a temporary project office by pooling all personnel and resources in order to streamline the operation.

Staff and training needs

Staff and training needs should be well identified during the study of institutional capabilities.

Since this will affect project implementation, recommendations should be made carefully. For example, unrealistic demands of staff increase will not be accepted by most governments, and bad timing for fellowships abroad could hamper project implementation. A network analysis of training needs including proper timing can be a plus for efficient implementation.

Farmers' training

Recommendations on farmers' training should include number of farmers or leaders to be trained, training subjects, timing and costs. Extension and education programmes for farmers in general should also be included. Demonstration plots to be established on public or private lands should be planned and their costs estimated.

Research needs

Research needs are usually included in the recommendations. For practical purposes, emphasis is normally laid on applied research for solving immediate problems and needs, leaving long-term basic research to regular research institutions.

Such applied research may include the following broad categories:

- Better resources management alternatives.

- Cost effective watershed conservation measures including new techniques introduced from outside.

- Practical monitoring and evaluation methodologies.

Pre-implementation work

Recommendations should be made on pre-implementation or pre-project work. Specific recommendations should include:

- Organizing or recruiting project personnel.

- Pre-project training of key staff using the existing budget.

- Initiation of extension activities with regular resources.

- Collecting further information for implementation.

- Preparation for equipment and vehicles procurement.

- Sub-contract preparation, if necessary.

- Other administrative arrangements.

This pre-implementation work is absolutely necessary if project implementation is to begin according to the schedule.

Work schedule and its control

Finally, a work schedule should be included as part of the recommendations. It can be expressed as a bar chart, a flow chart or a network analysis. The important thing is to consider progress logistics and to streamline project operations. All major activities should be scheduled in a sequence that avoids congestion and bottlenecks.

Each component (forestry, soil conservation, extension, infrastructure, etc.) will also draw up its own sub-schedule based on the master plan or schedule.

It is not enough to just set schedules. Means for controlling progress must also be developed. Recommendations should also be made on how:

- to ensure manpower supplies, e.g. technical staff, foreign experts, or short-term consultants in planned sequences;

- to obtain equipment, vehicles, or sub-contracts, etc. according to the work schedule;

- to establish channels to get high-level support on policy, finance, and administration;

- to build a system to oversee field work including inspection and ' reporting; and

- to set up an overall mechanism for monitoring and periodic evaluation of work quality and progress;

11.3 Monitoring, evaluation and follow-up

Monitoring and evaluation systems

A monitoring and evaluation system should be built into the project in order to permit periodic appraisal of the project's performance, physical outputs, benefits, expenditures and impacts.

Unfortunately, in the past, most watershed projects have been inadequately monitored and evaluated, and results were often poorly documented and disseminated. Many of the difficulties were due to lack of methodology and to the time and mechanisms needed for such activities. The Mechanisms needed by this kind of project may include the following:

- A data base and monitoring unit should be established within the project to collect, collate and analyse data for the use of evaluations. This can be done by using microcomputers (see Appendix 3).

- An independent evaluation body to undertake periodic appraisal work. Its members may be drawn from national planning agencies, universities, research institutes, interest groups and local communities, etc., in addition to project staff.

- A chapter in the project's annual and final reports on monitoring and evaluation results. Achievements should be clearly set out and compared to the original goals.

Independent evaluations should be carried on even after the project is completed. Because of the long-term effects of watershed work, these evaluations though difficult, are often useful. Depending upon the availability of data and resources, the whole or part of the watershed project should be evaluated periodically. The lack of information on long-term results is a major concern to many planners, government authorities and funding agencies.

Monitoring and evaluation methodology

The final plan or project proposal should include proper methodology on monitoring and evaluation. Although such work depends on watershed management objectives, the general methodology may include the following major indicators of a project:

For erosion and sediment reduction:

- Set a hydro-meteorological network to collect and monitor rainfall, streamflow, sediment and pollution data for long-term analysis and comparison.

- Make reservoir, pond or check dam profile surveys to obtain data on sedimentation rates and volumes.

- Establish small plots on major soils and cropping systems with and without conservation measures to monitor and evaluate differences in soil erosion and runoff.

For changes in land use and vegetative cover:

- Obtain aerial photographs or satellite remote sensing data and make studies on periodic changes (every 5 to 10 years).

- Conduct sampling surveys for special purposes or for needed information.

For flood prevention benefits:

- Establish rainfall and stream gauging stations as mentioned above.

- After major storms and floods, survey damages to compare with predictions and past events.

For farm improvement benefits:

- Keep some farm records from selected farms for monitoring purpose.

- Make periodic farm management surveys for comparison with the baseline survey data.

- Conduct specially designed surveys on farm production and income.

For other socio-economic benefits:

- Repeat the baseline socio-economic surveys every 5 to 10 years to compare the results.

- Make special surveys, if needed.

As mentioned previously, the unit which is responsible for establishing the data base and for routine monitoring should assist in the various phases of evaluation work. A personal or microcomputer will help to facilitate data storage, analysis and comparison (see Appendix 3).

Follow-up

Watershed planning cannot be considered complete if the project document ends up in a filing cabinet or on a bookshelf. Whoever is responsible for planning should follow it up, to see that the project is properly financed, either by the government or by international agencies, and approved for action.

Finally, watershed managers and planners should realize that planning is a continuous effort. In many countries, original planners are also required to be involved in project implementation, monitoring and evaluation. Experience thus gained can be used for the planning of similar projects in the future.

Appendix 1

SAMPLE CHECKLIST FOR WATERSHED SURVEY II

I. Watershed Name and Location

A. Name of watershed

B. Location (County or counties and State. For small watersheds it may be desirable to give in addition the legal description)

Physiographic region

Tributary to

C. Size: area in acres or hectares

area in square miles or square kilometres

II. Watershed Characteristics

A. Climate (can get climatic data from Weather Bureau)

1. Precipitation (in inches or mm)

a) Total annual
b) Seasonal (several breakdowns possible here¯indicate months)
c) Form (snow, rain¨give percents of each if available) d) Maximum precipitation intensities (if available)

2. Evaporation (annual and seasonal if available)

3. Wind a) Prevailing wind direction b) Wind hazard high, moderate, or low

4. Other pertinent climate data such as relative humidity, climatic type, etc.

B. Geology and physiography

1. Size and shape

a) Area in acres or hectares Area in square miles or square kilometres
b) General shape of watershed

2. Elevation in feet or metres: specify locations

a) At high point on divide
b) At mouth or gaging station
c) At headwaters of permanent stream
d) Fall of stream in feet per mile or metre per kilometre

3. Slope and aspect

a) Slope - proportion of watershed in different slope classes
b) General orientation or aspect of watershed

4. Drainage features

a) Drainage pattern (dendritic, radial, annular, etc.)

b) Drainage density

1) Number of well-defined channels per square miles or kilometres of watershed
2) Miles or kilometres of permanent stream per square mile or kilometre of watershed

c) Location and description of lakes, bogs and swamps, if present

5. Parent rock-igneous, metamorphic, sedimentary

a) Percent of each and specific kind
b) Condition of parent rock: solid, fractured, faulted, extent of outcrops.

C. Soils

1. Proportion of watershed area in:

a) Residual soils
b) Glacial material (specific type)
c) Alluvium
d) Volcanic materials

2. Distribution and hydrologic characteristics of major soil groups in watershed:

a) Depths: shallow, medium or deep
b) Infiltration capacities (if available) texture classes give an indication
c) Depth of confining layers, if present
d) Surface drainage conditions
e) Erosiveness of soil

D. Land use and cover conditions

1. Distribution by use classes: forest, range, agriculture, urban, metropolitan, etc.

2. ownership pattern

a) Public, private, industrial
b) Stability of ownership

3. Forest land conditions (include fire history and past use where applicable)

a) List major forest types, their use and condition

1) Good--good stocking, light or no grazing, good litter cover, no evidence or erosion, etc.
2) Medium¯moderate stocking, no overgrazing or excessive compaction, thin but continuous litter cover, no marked evidence of erosion,
3) Poor¯inadequate stocking, soil compacted heavy to overgrazing, thin or partial litter cover, sheet erosion or gullies present

4. Range land¯indicate condition based upon evidence of overgrazing, type and condition of cover, soil compaction, pedestaling, erosion pavement, gullies, etc.

5. Agricultural land

a) Major agricultural economy (i.e., dairying, truck farming, etc.)
b) Improper locations (steep, thin soils, etc.)
c) Improper practices (up and down hill crops, clean cultivation, overgrazing in both pastures and farm woodlots)
d) Amount of cropland under irrigation
e) Agricultural drainage
f) Extent to which conservation practices are applied

6. Developments

a) Urban and/or metropolitan areas within watershed
b) Roads
1) Road network: sparse, dense, absent
2) Road conditions: primary and secondary roads

7. Other uses

a) Intensity of recreational use

1) Resort use

2) Wildlife resource

3) Fish resource

III. Watershed hydrology

A. Erosion conditions along stream

B. Floods

1. Do floods occur frequently, periodically, seldom?

2. Season of year when most floods occur

3. Cause: high intensity storms, spring break-up, etc.

4. Maximum stage or heights

5. How much flood damage has resulted?

C. Stream flow

1. Quantity

a) Source of streamflow (lakes, bogs, springs, ground water flow, accumulated snow, etc.)

b) Annual yield (in c.f.s. or c.m.s. and as percent of precipitation)

c) Seasonal yield

d) Maximum and minimum yields

e) Flow regime¨do streams rise rapidly after rain, get low, or disappear in dry periods?

2. Quality

a) Do streams generally flow clear, turbid, turbid only in floods or high water stages?

b) Is water of poor quality or polluted?

1) Natural pollution¨organic, drainage from swamp and bog areas

2) Industrial

3) Municipal

4) Agricultural

5) Hardness, mineral content

IV. Water uses and needs

A. Source(s) of present supply surface water ground water

B. Domestic

C. Irrigation

D. Industrial

E. Power generation F. Recreation and wildlife

G. Other (e.g., sewage treatment)

H. Are present supplies adequate?

I. Will present supplies meet projected needs?

V. Water Problems

A. Erosion

B. Flooding (including siltation)

C. Water supply

D. Water quality

E. Effects of land use on the problem(s)

1. Accelerated surface runoff, erosion, flooding

2. Water pollution

3. Reduced recreational values

4. Reduced standard of living

5. Unstable community

Source: Colorado State University.

Appendix 2

NEW SCHEME OF LAND CAPABILITY CLASSIFICATION [1]

A Treatment-Oriented Scheme especially for Hilly Watersheds

Slope / Soil depth	1. Gently sloping $< 7^\circ$	2. Moderately sloping 7°–15°	3. Strongly sloping 15°–20°	4. Very strongly sloping 20°–25°	5. Steep 25°–30°	6. Very steep $> 30^\circ$
Deep (D) >36 in (> 90 cm)	C_1	C_2	C_3	C_4	FT	F
Moderately deep (MD) 20-36 in (50–90 cm)	C_1	C_2	C_3	C_4 / P	FT / F	F
Shallow (S) 8–20 in (20–50 cm)	C_1	C_2 / P	C_3 / P	P	F	F
Very shallow (VS) <8 in (< 20 cm)	C_1 / P	P	P	P	F	F

1 Symbols for most intensive tillage or uses:

C_1 : Cultivable land 1, up to 7° slope, requiring no, or few intensive conservation measures, e.g. contour cultivation, strip cropping, vegetative barriers, rock barriers and in larger farms, broadbase terraces.

C_2 : Cultivable land 2, on slopes between 7° and 15°, with moderately deep soils needing more intensive conservation, e.g. bench terracing, hexagon, convertible terracing for the convenience of four-wheel tractor farming. The conservation treatments can be done by medium-sized machines such as Bulldozer D5 or D6.

C_3 : Cultivable land 3, 15° to 20°, needing bench terracing, hexagons and convertible terracing on deep soil and hillside ditching, individual basin on less deep soils. Mechanization is limited to small tractor or walking tractor because of the steepness of the slope. Terracing can be done by a smaller tractor with 8 ft (2.5 m) wide blade.

C_4 : Cultivable land 4, 20° to 25°, all the necessary treatments are likely to be done by manual labour. Cultivation is to be practised by walking tractor and hand labour.

P : Pasture, improved and managed. Where the slope is approaching 25°, and when the land is to wet, zero grazing should be practised. Rotational grazing is recommended for all kinds of slopes.

FT : For food trees or fruit trees. On slopes of 25° to 30°, orchard terracing is the main treatment supplemented with contour planting, diversion ditching and mulching. Because of steepness of the slopes, interspaces should be kept in permanent grass cover.

F : Forest land, slopes over 30°, or 25° to 30° where the soil is too shallow for any of the above mentioned conservation structures.

1. Slopes and the number of slope classes can be modified to meet country's needs and some F lands between 25° to 30° can be used for agroforestry purpose.

2. Any land which is too wet, occasionally flooded or too stony, which prevents tillage and treatment should be classified as: (a) below 25°: pasture; (b) above 25°: forest.

3. Gully dissected lands which prevent normal tillage activities: forest (over 25°) or pasture below 25°,.

4. Mapping Symbols: It could be labelled as follows:

$$\frac{\text{Most intensive use}}{\text{soil - slope - depth}}$$

example: $\frac{\text{C2}}{\text{32 - 2 - D}}$

means: $\frac{\text{Cultivable Land 2}}{\text{Wirefence Clay Loam - 7° to 15° - 36 in(90 cm)}}$

Or, it could be simple labelled as C_2.

REMARKS:

(a) Slope classification

Slopes are divided into six categories, each having its implications for conservation treatments and the kind of tools to be used:

< 7° Flat to gently sloping. Broadbase terraces or other simple conservation treatments can be used up to 7°. Full mechanization for cultivation is applicable in this category. This slope class may not be common in hilly watersheds.

7°-15° Moderately sloping. Medium sized machines such as a Bulldozer D5 or D6 can be employed for bench terracing. Four wheel tractor mechanization for cultivation can be applied.

15°-20° Strongly sloping. Small sized machines such as b4 can be employed for conservation treatments. Small tractors, or walking tractors can be used for cultivation.

20°-25° Very strongly sloping. Manual for building the structures Hand labour and walking tractor for cultivation.

25°-30° Steep. Only for permanent tree crops such as food trees, fruit trees, forest or agroforestry. Manual labour for treatments.

>30° Very steep. Needs forest cover.

(b) Soil depth

Soil depth is divided into four classes. Here the depth refers to the effective depth of the soil which machine or manual labour can cut for conservation treatments and which plant roots can penetrate.

< 8 *in* Very shallow. Only on nearly level land can cultivation be (20 cm) practised.

8-20 in Shallow. Only below 20° slopes can this be cultivated with (20-50 cm) conservation treatments.

20-36 in Moderately deep. On a 25° slope, *for* instance, it needs about (50-90 cm) 30 *in* (76 cm) of soil to make narrow terraces of 8 ft (2.5 m) wide.

> 36 in Deep. No further soil depth classification is needed because (90 cm) the riser or terrace is limited to 6 ft height which is 3 ft cut and 3 ft fill.

(c) Other limiting factors

Land which is too wet, has poor drainage, occasionally floods or is too stony; which permanently limits the tillage or treatment, should be classified for lower or less intensive uses. On slopes under 25° such land can be used as pasture, whereas on slopes over 25° forest cover is proper so far as erosion control is concerned. Gully-dissected land which prevents any tillage activity should be put under permanent cover.

(d) Capability classes

Land is classified into its most intensive tillage or use. T ere are four major classes - cultivable land, pasture, food trees and forest . Only cultivable land has four sub-classes, each having implications for needed conservation treatments and tools to be employed. Use according to or within the capability class is encouraged, whereas use beyond the capability class is discouraged.

(e) Soil conservation treatment

In addition to the most popular conservation treatments on gentle slopes (below 7°) such as broadbase terraces and strip cropping, etc., six major treatments for steeper slopes are taken into account for the basis of the new classification scheme. These six treatments, which have been established in the hill slopes of Taiwan as well as in the western part of Jamaica under the UNDP/FAO project JAM/67/505, are particularly suited for the humid tropics. Bench terraces, hillside ditches and individual basins can be used to treat slopes up to 25° if the soils are deep enough. Orchard terracing can be applied from 25° to 30° slope. Convertible terracing and hexagons for full mechanization are to be employed on slopes up to 20°. All of them are mainly reverse sloped terraces of varying widths. Later another type of terracing was added: Intermittent terraces (see FAO Conservation Guide 13/3).

Appendix 3

USE OF MICROCOMPUTERS: AN INTRODUCTION

1. Microcomputers as a new tool

Microcomputers are now inexpensive and affordable in many developing countries. They can handle large quantities of watershed data, and may conveniently store, edit, retrieve, combine, update, and analyse them when needed. Many commercial software packages are available for various applications. Watershed managers worldwide are increasingly using microcomputers as a new tool to assist them in watershed survey and planning.

In addition to establishing a watershed database, mapping can be carried out more easily and rapidly with proper computer hardware and software. It is especially convenient when several maps are overlaid and manipulated to generate new spatial information. From an established database, detailed descriptive information can be combined with or extracted from maps readily.

This chapter will give a brief introduction to the subject. It will not explain the technical details of how computers and software are used, which would need a separate manual. Instead, this chapter will provide the planners and watershed managers some general guidelines and current information on using microcomputers for survey and planning.

2. Establishment and management of watershed data

Basic hardware for data management

Depending on the volume of data to be managed and whether the microcomputer is solely used for this purpose, the following suggestion is for average conditions (i.e. a watershed around 50 000 ha populated with 5 000 to 10 000 farmers). If the areas are bigger and farmers are more, branch offices would normally be set at sub-watershed level and each of them could have a set of equipment. The actual needs for hardware and software of a project, however, should be determined in consultation with an experienced computer expert.

- A microcomputer, IBM PC/AT or compatible with 640 kilobytes (KB) of RAM core storage, two floppy disk drives, and a hard disk of 40 Megabytes (MB).

- A 13 inch colour or monochrome monitor with compatible video card.

- A near letter quality printer, 9 to 24 pin dot-matrix type.

* Brands mentioned in this chapter are for reference only and do not constitute FAO's endorsement.

The most favourable purchase price for this set is around US$ 2 500. However many products with similar specifications may be available and it would often be worthwhile considering several alternatives before buying. Furthermore it is to be expected that prices for computer hardware will go down rather than up in the next five years.

Basic software for data management

There are now many software packages which are suitable and new software packages are constantly being developed which may be even better and cheaper than the existing ones. The following software are among the basic and popular ones which have been widely used or tested:

- *DOS :* This is a Disk Operating System. Its main functions are for managing disks and disk files. DOS also gives the operator complete control over the computer. It links between the operator and the computer. MS-DOS, produced by Microsoft Corporation is a system applicable to 50 leading microcomputers.

- *Word Processors:* These are convenient for preparing, editing, and presenting reports, Many commercial word processing software packages are available such as Word Perfect, Word Star, Multimate, and Wordmarc, etc. A careful selection is required to fit the actual needs.

- *Spreadsheets*:_A spreadsheet is a powerful data analysis tool. It carries mathematical, statistical, and economic analysis functions, as well as database management and basic graphics. Lotus 123 is a spreadsheet that is widely used for technical analysis and modelling. It has a worksheet consisting of 256 columns and 8192 rows. The cells can be easily rearranged and edited. Cells can be built with dynamic formulas that refer to the contents of other cells. The data can be managed as a database, and can be transformed, modified, summarized and graphed. The possible applications are almost endless. For watershed survey and planning, spreadsheet models can be used for recording, *analysing,* retrieving, editing and monitoring physical data such as hydrology, land use and land capability, and socio-economic data as well. It can also be used for economic assessment, water balance estimation, and calculation of soil conservation treatment specifications.

- *Database Management:*_A database, different from a spreadsheet, is an organized record. A database file consists of records and each of them has a series of fields. A field is the area that contains a particular item of information. Database management systems (DBMS) are specialized tools for storing information in a systematic order that allow the user to retrieve, extract, sort and synthesize them in a large data sets. dBASE IV (a relational DBMS) and PFS (Professional File System: a flat file DBMS) are some of the software used extensively. The latter is a simple file management programme for project managers and professionals. dBASE IV can be used for individual *conservation* farm planning records and also for keeping farmer's progress of work, cost and government contributions, etc. in a multi-file operation setting.

Special software for watershed modelling and soil conservation

- Watershed models: Watershed modelling is a growing field and it basically uses computers. In the past, models were only applied on expensive mainframe computers which have limited use in developing countries. Only recently has their applications on microcomputers become a reality. This will undoubtedly open new opportunities in many developing countries to verify, modify and adopt existing models or create new ones for planning and assessment uses. Many models have been developed over the last decade. The following list contains some selected ones having microcomputer applications:

1) AGNPS-PC (Agricultural Nonpoint Source Pollution Model) is a watershed model for estimating erosion, sediment transport, and nutrient loading in surface water. Agricultural Research Service, USDA, Minnesota Pollution Control Agency and Soil Conservation Service.

2) ANSWERS (Areal Nonpoint Source Watershed Environment Response Simulation) is a model to provide estimates of runoff, erosion and sedimentation for small agricultural watersheds (10 000 ha). Department of Agriculture Engineering, Purdue University.

3) CREAMS (Chemical, Runoff and Erosion in Agricultural Management Systems) is a field data model for assessing the chemicals, runoff and soil loss from agricultural practices. USDA-ARS, Tifton, Georgia.

4) PRMS (Precipitation-runoff Modelling System) is similar to ANSWERS but tracks additionally snowmelt, soil moisture, evapotranspiration and percolation during inter-storm periods to perform long-term water balance calculations. US Geological Survey, Lakewood, Colorado.

5) SWRRB (Simulator for Water Resources in Rural Basins) is to predict management effect on water and sediment yields in ungauged rural basins. USDA-ARS, Temple, Texas.

- Erosion, conservation planning and evaluation: Software packages for predicting erosion are many; mostly using USLE (Universal Soil Loss Equation) as a base. Software for soil conservation planning And evaluation is still limited the following gives examples of some available software:

1) ICE (Interactive Conservation Evaluation) is a program that combines Universal Soil Loss Equation (USLE) calculations and basic economic analysis for evaluating cost effectiveness of alternative conservation measures. USDA-Soil Conservation Service, Fort Worth, Texas.

2) LANDCONS is an expert computer system for land capability classification and conservation farm planning for small farmers on steep watershed slopes. Computer Assisted Development Inc. (CADI) and Colorado State University, Fort Collins, Colorado.

3) SP (Simple Process) Model is a surface runoff and sediment yield model for single rain storms over a homogeneous surface. USDA-ARS, Fort Collins, Colorado.

4) USLE and MUSLE are famous models used for erosion prediction for long-term average soil losses. USDA-ARS or USDA-SCS.

The above-listed software packages are either public domain or can be acquired with reasonable prices other erosion and watershed models which may be useful include:

- Silsoe Model (an erosion model) National College of Agricultural Engineering, Silsoe, Bedford, Great Britain;

- SMAP (Soil Moisture Accounting Procedure Model: modified version) University of Sao Paulo Brazil;

- Stanford Model (a watershed hydrology model.) Stanford University, California;

- WASED (a small forest watershed model) Rocky Mountain Forest and Range Experiment Station, Flagstaff, Arizona.

3. For watershed mapping and related information systems

Geographic Information Systems (GIS)

A Geographic Information Systems (GIS) is a set of computer programs which input, store, analyse, and display spatial and non-spatial data (CSU 1989). A modern GIS may be viewed as an integrated system composed of five main compartments: data input, database, model base, decision support system, and information display (Loran et al. 1988). GIS differs from computer-aided mapping (CAM) because the former is analysis oriented while the later is display oriented. In simple term, GIS has the capability of mapping with a full range of information interaction.

The major advantages of using GIS for survey and planning can be summarized as follows:

- they allow for a variety of manipulation including map measurement, map overlay, transformation, graphic design, and database management;

- stored spatial and descriptive data can be retrieved quickly;

- rapid and repeated analytic testing of conceptual models about geography can be performed i.e. for land capability and land suitability systems;

- comparisons of spatial information over time can also be efficiency performed, i.e. monitoring land use changes;

- certain analysis which can not be performed cost effectively by manual methods can be done by GIS i.e. terrain models, aspects;

- a rapidly growing field as it will further integrate with other data analytical tools.

GIS inputs of spatial data can be a) maps, b) aerial photos, c) special digital data such as satellite imagery, and d) textual and numerical data. The basic system outputs are: a) maps, b) graphics, and c) text including statistical tables, computed data files, etc. The entire system and its process can be seen from a simple diagram as shown in Fig. 12.

Before acquisition or development of such a system, planners and watershed managers need to consider thoroughly the following conditions.

1) Selection of a proper system to meet the requirement.

2) Consideration of the cost of initial acquisition.

3) The use of the system, whether limited to the present planning and monitoring work, or to be used for other projects or additional work.

4) Cost and techniques of converting existing data into a digital file i.e. digitizing, scanning, data conversion.

5) Proper maintenance including computer upkeep, skilled technicians, and software maintenance. This may require a special administrative unit and a regular annual budget.

Fig. 12

GEOGRAPHIC INFORMATION SYSTEMS (GIS)

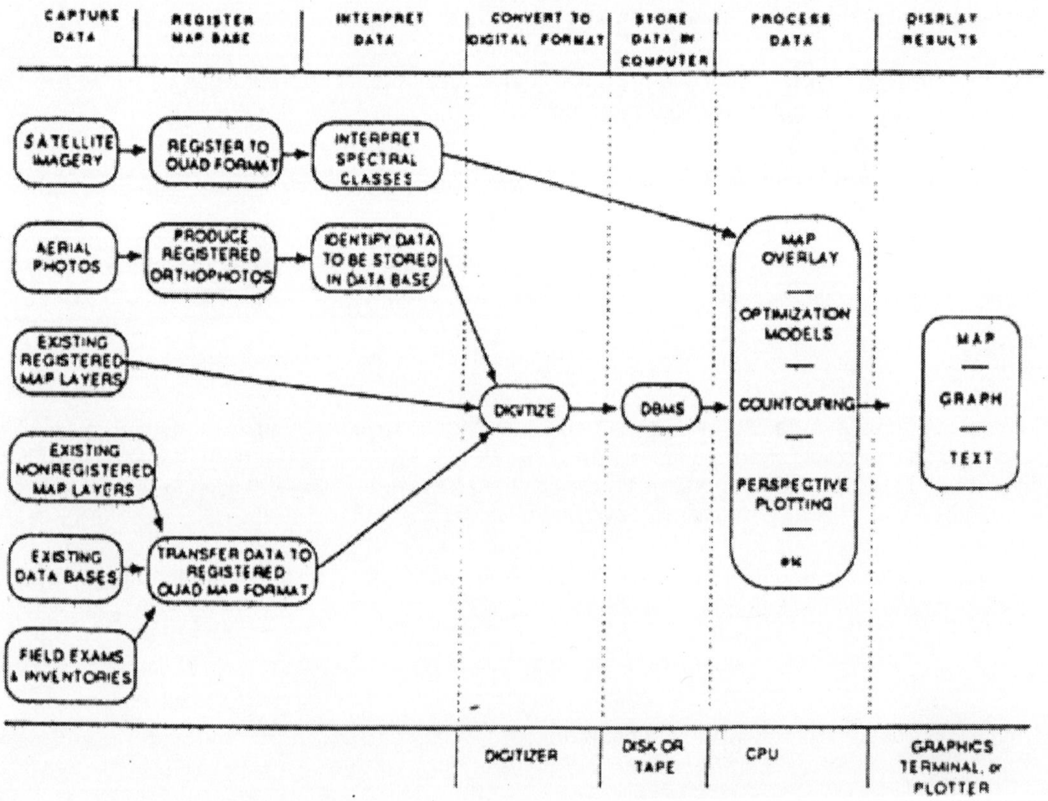

Source: Berry, 1988

GIS hardware and, software

Actual hardware and software needs for a project depend on objectives and availability of funds. Generally, in addition to a host computer, spatial data entry and display require the following basic hardware:

- a digitizer for map input;

- a digital tape drive if satellite image processing is involved;

- a colour ink plotter for map production;

- a mouse for screen manipulation and editing.

For the host computer, current estimates suggest that a 80386 microcomputer with 1.2 megabyte (MB) of internal memory supported by disk storage of some 100 MB or more are sufficient for many purposes. If it is insufficient for handling large data sets or for analysis-intensive projects, a 32 bit minicomputer should be acquired. Or, purchase a workstation with several microcomputers to share the work.

Hardware cost varies greatly depending on the type, source, sophistication, and the supporting equipment chosen (i.e. large plotters and digitizers are much more expensive than small ones). A moderate set of hardware including a host microcomputer, a colour monitor, a digitizer, a tape drive and a plotter cost around US$ 25 000 at 1589 price. A similarly configured minicomputer-based hardware will cost an additional US$ 25 000 to 40 000. A workstation will cost about US$ 4 000. GIS software may range from several hundred to ten thousand US Dollars and more. For example, the popular Earth Resources Data Analysis Systems (ERDAS) combining image processing capability and raster GIS, cost about US$ 25 000 for the basic hardware and software. With expanded hardware and software, the prices can easily reach US$ 50 000. Its GIS software module and Image Processing software module alone costs US$ 4 000 each. The popular PC Arc/Info system is a vector type GIS with topological structure and a relational database. The complete PC software package sells for about US$ 13 000. These prices, however, are still. affordable in many developing countries.

There are also some inexpensive GIS software packages. For instance, pMap (Raster system) costs about US$ 900, MapInfo (Vector system) costs US$ 750, and WOW (Both raster and vector), US$ 500 (all 1989 prices). These are systems all used on microcomputers. However, they have their usage under a specific set of conditions or have their special functions. The users should understand their advantages and disadvantages before acquiring them.

Recently, the International Institute for Aerospace Survey and Earth Sciences (ITC) of the Netherlands has designed a special GIS system called Integrated Land and Watershed Management Information System (ILWIS) which integrates many computer-based models (erosion, socio-economic. etc.) with map analysis and remote sensing systems.

4. A future outlook

Using microcomputers together with GIS, watershed models, and database programs for watershed survey and planning is a new trend. Microcomputer hardware and software will both be cheaper in the foreseeable future. Training opportunities for these subjects are increasing rapidly through regular school curricula or short courses. Because there are many advantages of using microcomputers and their costs are reasonable, some developing countries are starting to use them for watershed survey and monitoring work with good results.

REFERENCES

Berry, J.K. 1988. Management, Mapping and Analysis for Geographic Information Systems (GIS). Lecture notes for GIS short course, Colorado State University, Fort Collins, Colorado.

Bochet, J. 1983. Management of Upland Watersheds: Participation of the Mountain Communities. FAO Conservation Guide No. 8. Rome.

Brooks, K.N. & H.M. Gregersen. 1985. The Economics of Watershed Management¨ Problems and Recommendations for Project Analysis. Paper presented to FAO/ICIMOD/East¨West Center "Expert Meeting on Strategies, Approaches, and Systems for Integrated Watershed Management". Kathmandu, Nepal.

Chang, M. 1982. Laboratory Notes: Forest Hydrology. Stephen F. Austin State University. Nacogdoches, Texas.

Colorado State University. 1989. Introduction to GIS (Lecture Notes). Cooperative GIS Technology Laboratory. Fort Collins, Colorado. Dangermond, J. 1988. A Classification of Software Components Commonly Used in GIS. Lecture notes for GIS short course. Colorado State University, Fort Collins, Colorado.

de Graaff, J. & Q. Laumans. 1981. Economic Aspects of Watershed Development. FAO/UNDP JAM/78/006 Project Working Paper 5. Kingston, Jamaica. Dixon, J.A. & K.W. Easter. 1985. Economic Aspects of Integrated Watershed Management. Paper presented to the "Workshop on Integrated Watershed Management". East-West Center, Honolulu, Hawaii.

Easter, K.W. & C.J.N. Gibbs. 1985. Implementation and Institutional Aspects of Integrated Watershed Management. Paper presented to the "Workshop on Integrated Watershed Management". East¨West Center, Honolulu, Hawaii.

Easter, K.W. & M.M. Hufschmidt. 1985. Research for Integrated Watershed Management in Developing Countries. (Draft paper). East¨West Center, Honolulu, Hawaii.

Eren, T. 1972. Project Formulation in Watershed Management. In FAO Report No. TA 3112. Rome.

FAO. 1976¨89. FAO Conservation Guide 1, 2, 3, 4, 8, 13/3 and 13/ 4. Rome. FAO. 1976a. A Framework for Land Evaluation. Soil Bulletin 32.

FAO. 1976b. Comprehensive Integrated Watershed Development. Paper W/K 0653. Rome.

FAO. 1977a. Project for the Rehabilitation and Development of the Pindars River and Two Meetings Watershed (Jamaica). JAM/67/505 Project Technical Report 13/1. Rome.

FAO. 1977b. Farm Management Data Collection and Analysis. FAO Agricultural Service Bulletin No. 34. Rome.
FAO. 1979. Watershed Development, with Special Reference to Soil and Water Conservation. FAO Soils Bulletin 44. Rome.

FAO. 1980. Economic Analysis of Forestry Projects (Readings and Case Studies). FAO Forestry Paper 17. Rome.

FAO. 1983a. 7th UN/FAO International Training Course in Remote Sensing: Applications to Thematic Mapping with Special Reference to Land Use. RSC Series 20. Rome.

FAO. 1983b. Guidelines: Land Evaluation for Rainfed Agriculture.

FAO. 1984a. Report on the Training Course on Watershed Management for Africa. GCP/RAF/179/FIN. Rome.

FAO. 19$4b. Land Evaluation for Forestry.

FAO. 1985. Guidelines: Land Evaluation for Irrigated Agriculture. FAO. 1987. Guidelines for Economic Appraisal of Watershed Management Projects. Conservation Guide 16. Rome.

Ffolliot, P.F. 1985. Multiple Use Implications of Watershed Management. Paper presented for the "Workshop on Integrated Watershed Management". East-West Center, Honolulu, Hawaii.

Ffolliot, P.F. & J.L. Thames. 1983. Environmentally Sound Small-Scale Forestry Project, Guidelines for Planning. Vita Publications, Arlington, Virginia.

Fleming, W.M. 1978. Phewa Tal Watershed Management Proposal. FAO NEP/74/020 Project Technical Report No. 5. Rome.

Fulcher, G.D. 1972. Economic and Institutional Aspects of Watershed Management. In FAO Report No. TA 3112. Rome.

Gittinger, J.P. 1972. Economic Analysis of Agricultural Projects. Johns Hopkins. Baltimore, MD.

Gregersen, H.M. & K.N. Brooks. 1980. Economic Analysis of Watershed Projects: Special Problems and Examples. In FAO Forestry Paper 17. Suppl. 2, Rome.

Hamilton, L.S. & P.N. King. 1983. Tropical Forested Watersheds, Hydrologic and Soils Response to Major Uses Or Conversion. Westview Press, Boulder, Colorado.

Hill, I.D. 1982. Natural Resources Surveys in Agricultural Development Planning: A Quick and Clean Method. In Agricultural Administration Vol. 1Q. Applied Science Pub. Ltd. England.

Hudson, N.W. 1981. Soil Conservation. Cornell University Press. Ithaca, New York, and Batsford, London, England.

Hufschmidt, M.M., D.E. James, et al. 1983. Environment, Natural Systems, and Development. An Economic Valuation Guide. Johns Hopkins Baltimore, MD.

Kunkle, S., W.S. Johnson & M. Flora. 1987. Monitoring Stream Water for Land-use Impacts: A Training Manual. National Park Service, Port Collins, Colorado.

Lal, R. 1988. Soil Erosion Research Methods. (Edited). Soil and Water Conservation Society. Ankeny, Iowa.

Loran et al. 1987. Management Conservation and Erosion Data Base. Paper presented to 5th Inter. Soil Conservation Conference, Bangkok. Muthoo, M.K. 1976. Economic Evaluation of the Environmental Effects of Erosion. Agriculture and Environment 3. Elsevier Scien. Pub. Co. Amsterdam.

OECD. 1986. The Public Management of Forestry Projects. Paris.

Pereira, C. 1985. The Management of Tropical Watersheds. Paper presented to the

FAO/ICIMOD/East-West Center "Expert Meeting on Strategies, Approaches and Systems for Integrated Watershed Management". Kathmandu, Nepal.

Sfeir-Younis, A. 1983. Economic Aspects of Soil Conservation Programs in Less-Developed Countries (LDCs). In Water International vol. 8. Elsevier Sequoia, Netherlands.

Sfeir-Younis, A. 1985. The Management of Sediments in Developing Countries: A Socioeconomic Perspective. Paper presented to the "Workshop on Integrated Watershed Management". East-West Center, Honolulu, Hawaii.

Shaner, W.W. 1979. Project Planning for Developing Economics. Praeger Publishers. New York.

Shaner, W.W., R.F. Philipp, et al. 1982. Farming Systems Research and Development: Guidelines for Developing Countries. Westview Press, Boulder, Colorado.

Sheng, T.C. 1966. Landslide Classification and Studies in Taiwan. JCRR, Taipei, Taiwan.

Sheng, T.C. 1972. A Treatment-oriented Land Capability Classification Scheme: For Hilly Marginal Lands in the Humid Tropics. In FAO Report No. TA 3112. Rome.

Sheng, T.C. 1981. Physical Planning for Watersheds. FAO/UNDP JAM/78/006 Project Working Paper 17. Kingston, Jamaica.

Sheng, T.C. 1985. Watershed Management Planning: Practical Approaches. Paper presented to the FAO/ICIMOD/East-West Center "Expert meeting on Strategies, Approaches and Systems for Integrated Watershed Management". Kathmandu, Nepal.

Sheng, T.C. & H.R. Stennett. 1975. Lecture Notes: Watershed Management and Soil Conservation Training Course. UNDP/FAO JAM 505 Project, Kingston, Jamaica.

Spears, J.S. & R.D.H. Rowe. 1980. Preliminary Guidelines for Designing Watershed Rehabilitation Projects. World Bank. Washington, D.C. Striffler, W.D. 1979. Watershed Planning and Management. In Planning the Uses and Management of Land. American Society of Agronomy, etc. Madison, WI.

Trimble, Jr. et al. 1957. How Far From a Stream Should a Logging Road be Located? Journal of Forestry, Vol. 55, No. 5.

Tolley, G.S. & F.E. Riggs. 1961. Economics of Watershed Planning. Iowa State University Press. Ames, Iowa.

Unesco. Man and Biosphere (MAB) Program. 1980. Watershed Resources Management and Environment Monitoring. Regional Training Course. U.S. AID/MAB, Washington, D.C.

UNIDO. 1972. Guidelines for Project Evaluation. UN, New York.

U.S. Forest Service. 1958. Land Treatment Measures Handbook. 2533. Washington, D.C.

U.S. Geological Survey. 1960. Flood-Frequency Analyses. Water Supply Paper 1543-A. Washington, D.C.

U.S. Geological Survey. 1973. Regional Analyses of Streamflow Characteristics. Government Printing Office, Washington, D.C. U.S. Soil Conservation Service. 1964. Economics Guide for Watershed Protection and Flood Prevention. Washington, D.C.

U.S. Soil Conservation Service. 1972. Engineering Handbook. Section 4: Hydrology. Washington, D.C.

U.S. Soil Conservation Service. 1983. National Soils Handbook. Washington, D.C.